T0275901

AIR: THE EXCELLENT CANOPY

ABOUT THE AUTHOR

Frank Fahy

Frank Fahy is Emeritus Professor of Engineering Acoustics at the Institute of Sound and Vibration Research (ISVR) of the University of Southampton in the UK. He began his career in the late 1950s as an aeronautical engineer designing and testing dynamic transonic wind tunnel models of aircraft such as Concorde. He joined the ISVR at its inception in 1963 as British Aircraft Corporation research fellow working on interior noise in aircraft caused by turbulent boundary layer flow over the fuselage. He was subsequently awarded a PhD for research into theoretical and experimental modelling of acoustically induced vibration of gas-cooled nuclear reactor gas circuit structures and was supported for 13 years by the national electricity generating company in running a research team working on the prediction and control of problems of audio frequency structural vibration and dynamic stress in this area of reactor design and testing.

Professor Fahy played an important role in the development and standardisation of means of measuring the magnitude and direction of energy flow in sound fields (sound intensity) which has greatly improved our ability to locate, distinguish and quantify individual sources of noise in the presence of other sources, such as in factories. He has researched many other areas of sound and vibration including low frequency noise transmission in buildings and ships, the modelling of sound fields in large factory spaces, violin acoustics, sound propagation in liquid-filled pipes, propeller noise in aircraft, and optimisation of rocket launcher nose fairings to protect satellite structures from launch noise. He has acted as consultant on the acoustic design of over twenty theatres and multi-purpose auditoria, on problems of noise, vibration and fatigue in petrochemical plant, and to many industrial companies on diverse vibroacoustic problems.

For outstanding contributions to the field of acoustics, Professor Fahy has been awarded the Tyndall silver and Rayleigh gold medals of the Institute of Acoustics, of which he is an honorary fellow, the Helmholtz Medaille of the German Acoustical Society (DEGA) for outstanding lifetime achievement, and the Medaille d'Etrangers of the French Acoustical Society (SFA) for Anglo-French collaboration.

He is the author of five textbooks and monographs on sound and vibration and co-editor of five others.

AIR: THE EXCELLENT CANOPY

Frank Fahy
Emeritus Professor, University of Southampton
United Kingdom

WOODHEAD
PUBLISHING

Oxford   Cambridge   Philadelphia   New Delhi

Published by Woodhead Publishing Limited,
80 High Street, Sawston, Cambridge CB22 3HJ
www.woodheadpublishing.com

Woodhead Publishing, 1518 Walnut Street, Suite 1100, Philadelphia,
PA 19102-3406, USA

Woodhead Publishing India Private Limited, G-2, Vardaan House, 7/28 Ansari Road,
Daryaganj, New Delhi – 110002, India
www.woodheadpublishingindia.com

First published by Horwood Publishing Limited, 2009
Reprinted by Woodhead Publishing Limited, 2011

British Library Cataloguing in Publication Data
A catalogue record for this book is available from the British Library

ISBN 978-1-904275-42-8

Cover design by Jim Wilkie.

This book is dedicated to the Womens Institute of King's Somborne, Hampshire, whose challenging invitation to give a talk about sound sparked off the idea of writing this book

Table of Contents

Author's Preface

This most excellent canopy, the air, look you, this brave o'erhanging firmament, this majestical roof fretted with golden fire

William Shakespeare 1601 Hamlet Act 2 Scene 2

You can't see it; yet the sky is blue. You can't touch it; yet you can feel its movement. It is very light and easily moved; yet it can support weights of hundreds of tons, destroy buildings and even move the Earth. It has no voice; yet conversation and music are impossible without it. It won't stop a bullet; yet it protects us from cosmic missile attack. It dries the washing; yet it brings us rain. It doesn't generate heat; yet it keeps the Earth, and us, from freezing. It is non-flammable; yet it allows us to make fire. It lacks life; yet it sustains it. These are a few of the multitude of attributes of the wonderful material that is 'Air'.

This book is a celebration of air: its origins, constitution, properties, behaviour, actions, functions and uses. It is not a formal scientific or technical account, but contains qualitative descriptions and explanations of physical phenomena and processes that are the concerns of science and technology, without recourse to theory or mathematical analysis. It also describes simple, cheap home experiments that illustrate many of the phenomena described in the text that could also be of value in school classroom science lessons. The principal aim of the author is to imbue the reader with an appreciation of the importance of this ubiquitous substance that most of us take for granted – until deprived of it!

Our atmosphere has suffered decades of chemical abuse that it has withstood in a quite remarkable fashion, but to which it is now succumbing. Although air does play a role in many of the processes contributing to global warming, it is far too complex and contentious a matter to be comprehensively and authoritatively covered in such a small book. However, mention is made of this threat to life on Earth at various points in the book.

A glossary of less common words and terms used in the book is presented, and a list of important references and suggestions for further reading is presented at the end.

Acknowledgements

I wish gratefully to acknowledge the patient efforts of John Nielsen-Gammon, Professor and Texas State Climatologist, Dept. of Atmospheric Sciences, Texas A&M University, Tamus, Texas, USA to instruct this argumentative author in the mysterious ways of the Coriolis effect. Any errors of fact or interpretation in this respect are entirely mine. I would also like to acknowledge with thanks the assistance with advice and graphic material received from John J Videler, Emeritus Professor of Evolutionary Mechanics of Leiden University and Leonardo da Vinci Professor of Marine Biology in Groningen University, The Netherlands. Discussions with Craig Gale concerning aerodynamic lift have been most valuable; he also brought Fig. 3.16 to my attention. In the search for copyright holders I have been generously assisted by Sarah Brooks, librarian of the Institute of Sound and Vibration Research, University of Southampton. I am grateful to my son, John, for carefully reading the draft and offering helpful suggestions for making the book more accessible to 'the man and woman in the street'.

Credits for the sources of figures reproduced from other publications are listed at the end of the book. I wish to offer my thanks to all those who generously made their material available to me, almost all at no cost. In spite of intensive efforts, it has either been impossible to identify the copyright holders of all the figures or to acknowledge owners of copyright who have not responded to our permission request. If you wish to claim copyright, please contact the publisher.

Glossary

Note: a number of the following words/terms have more than one definition. The ones given here relate specifically to the subject of the book

Ablation: melting and evaporation of material due to heating by friction

Acceleration: the rate of change of velocity (m/s^2)

Amplitude: the maximum deviation from the rest value of a quantity undergoing simple harmonic variation (see 'Simple harmonic')

Angle of incidence (attack): the angle between the direction of motion of an aerofoil relative to the undisturbed air (or angle between the direction of the airflow approaching an aerofoil) and a line from the leading to trailing edges of the aerofoil

Angular momentum (of a body about an axis about which it rotates): the sum over the whole elements of a body of the product of the mass of each element, its velocity and the perpendicular distance to the axis of rotation

Boundary layer: a thin layer of fluid that covers the surface of a solid body that moves relative to the fluid and in which viscous forces and turbulence are dominant features

Celsius (Swedish astronomer): scale of temperature in which water freezes at zero and boils at 100 under specified conditions

Centre of curvature: the centre of a circle that passes through a point on a curved line of any shape, has the same curvature and is tangential to the curve at that point

Coefficient: a factor that defines a proportional relationship of one variable quantity to another

Contiguous: touching

Convex: curved outwards like the exterior surface of a ball

Decibel: a logarithmic measure of the temporal variation of sound pressure divided by a reference sound pressure

Diffraction: a property of waves whereby their pattern of propagation in a supporting medium is altered by the presence of 'foreign' bodies in the medium

Diffusion: spreading in all directions (especially by interaction between randomly moving particles of a fluid)

Dimensions (physics): fundamental physical quantities in terms of which all others may be expressed, e.g., mass, length, time. For example, from Newton's Law, force has the dimensions of mass (M) times acceleration. Acceleration has the dimensions of velocity/time (LT^{-1}/T). So force has the dimensions MLT^{-2}

Element: a very small region of fluid

Empirical: resulting from, or relating to, observation and experiment, not to theory

Energy (mechanical): a physical quantity that is a measure of the capacity to do mechanical work (see 'Work)

Equilibrium: at rest or undeformed

Even/odd multiples: 2,4,6, 8, .../1,3,5,7,..... times
Fatigue (mechanical): damage to the basic structure of a material by repeated reversal of applied force/stress ultimately causing failure
Fluid: any non-particulate material that flows
Fluorocarbons: chemical compounds of fluorine and carbon
Fractional variation: the variation of a quantity divided by its equilibrium or time - average value
Frame of reference (rectangular Cartesian): a set of three mutually perpendicular axes (like three edges of a room that meet at a point) by means of which the position of any point in space may be defined by the perpendicular distances of the point from the three planes defined by the axes (e.g. distances to two joining walls and the floor)
Gradient: the rate of change of one physical quantity with change of another
Hand wing: the outer part of the bird wing which tends to be flatter in section than the inner 'arm wing' and has a shaper leading edge than the latter
Harmonic: (i) abbreviation for 'simple harmonic; (ii) a frequency component of a periodic signal. The fundamental frequency is the first harmonic and the frequencies of the higher harmonics are multiples of this frequency. The set of harmonics is known as a 'harmonic series'
Heat: a form of energy involving the random motion of molecules; also, the amount of energy in this form capable of doing work
Hydrodynamic: to do with the relation between forces and motions of fluids
Inversely proportional: proportional to unity divided by the value of a quantity
Kinetic energy: energy of motion
Latent heat: the heat absorbed or released in the process of change of phase (e.g. evaporation of water, condensation of steam)
Leading edge: the edge of a wing that an approaching flow meets first
Loudness: a subjective perception of the strength of a sound. A doubling of loudness correspond approximately to an increase of sound pressure level of 10 decibels (see 'Sound pressure level')
Micron: one millionth of a metre
Modulation: a time variation in the amplitude of an oscillation
Momentum: the sum over a body of the products of the mass of each element of the body and its associated velocity (vector quantity)
Natural frequency: the frequency with which a system continues to oscillate following the cessation of an action that disturbs it from a state of equilibrium
Newton: a unit of force which is roughly equal to the weight of a medium size apple
Newton's Laws of Motion: (i) a body continues in a state of rest or uniform velocity unless acted upon by a force; (ii) the acceleration of a body produced by a force is proportional to that force, in the direction of the force and inversely proportional to the mass of the body; (iii) action and reaction (forces) are equal and opposite (e.g. at an interface between contiguous bodies). Note: it would be more precise to specify the 'body' as a small particle and the concept of 'rest' implies the existence of an inertial frame of reference that undergoes zero acceleration. Although this cannot be defined, the assumption of its existence is often good enough to allow accurate analyses of motions to be made (see Chapter 5 and 'Coriolis')
Nonlinear: a relation between the variation of one quantity and that of another that cannot be expressed in terms of a constant factor

Overtones: similar to harmonics but not necessarily forming a harmonic series

Pantograph: the frame mounted on top of a vehicle that can be raised to contact a cable that carries the electric current that drives the vehicle

Parabolic relation: a proportional relation between the variation of one quantity and the square of another

Parameter: a measure of a physical system that characterises the system and/or its behaviour. It is usually non-dimensional meaning that is a pure number because it comprises a combination of physical quantities in such a way that the units cancel out. For example, the Strouhal number of an airflow across a cylinder that generates an oscillatory flow is given by St = frequency of oscillation (f) times cylinder diameter (d) over flow speed (U) in which the units cancel out. St is about 0.2 for such a flow. This means that the frequency of oscillation is given by f = 0.2 *U/d in all cases

Pascal: a unit of pressure which equals one Newton per square metre or approximately 145 millionths of a pound per square inch

Peripheral speed: the speed of the periphery of a rotating body

Permafrost: a state or region of the surface of the Earth of which the temperature is permanently below zero Celsius

Phase: a state of matter (e.g. liquid, gas, solid); a measure of the time difference between one harmonic oscillation and another of the same frequency

Potential energy: energy possessed by a body or system by virtue of its position (e.g. height above the ground) or configuration (e.g. energy stored in an archer's bow when bent). The latter form is also known as 'strain energy'

Proportionality (constant of): if a graph is drawn to show the relation between the variation of one physical quantity and another, and the graph is a straight line, the relation is said to be proportional (and linear). The slope of the graph gives the constant of proportionality

Refraction: the bending of the direction of wave propagation that occurs when a wave passes from one medium to another, or from one region of a medium to another region of the same medium, in which the speed of wave propagation is different

Resonance: the state of strong vibration of a system that occurs when the frequency of an exciting agent equals a natural frequency of the system

Reveals: those exposed areas of the frame of a double glazed window that bound the cavity between the panes

Scattering: redirection of the energy of an incident wave into many directions

Simple harmonic: variation of a quantity with time that is sinusoidal in form

Sinusoidal: taking the from of the sine function

Sound pressure level: a logarithmic measure of the variation of air pressure in a sound field. Its unit is the decibel. Doubling of the pressure variation corresponds to an increase of 6 dB

Speed: the rate of travel without regard to direction

Streamwise: following the direction of the local particle flow in non-turbulent flow

Tangential: in a direction defined by a straight line that touches but does not locally cross a curve

Trailing edge: the edge of a wing from which a flow leaves it

Trajectory: a curve that traces out the path of a moving body

Translation: movement in a straight line

Transverse: extending or proceeding cross-wise to the principal direction considered; lying or running across this direction

Vector: a quantity that possesses both magnitude and direction. Vectors are conventionally represented by arrows; the length represents magnitude and the direction represents direction

Velocity: the vector of which speed is the magnitude

Viscosity: the phenomenon that acts so as to resist relative 'sliding' movements of adjacent regions of fluid and also to resist movement of bodies through a fluid

Voice coil: a coil of conducting wire that is wound on a tubular form and employed in a loudspeaker

Voidless: having no empty space(s); continuous

Volume acceleration: the rate of change of the rate of volume flow of a fluid

Volume velocity: the rate of volume flow of a fluid

Vortex/vortices: rotational systems of flow

Vorticity: a measure of the rate of rotation of fluid particles

Wake: the region of de-energised flow downstream of a body in a fluid flow caused by the separation of the boundary layer(s) from the body

Work (mechanical): work is defined as force times the distance moved by the point of action of the force in the direction of the force vector. When work is done on a system, its energy increases; when a system does work, its energy decreases

1

Origin, Nature and Properties of Air

The earth is surrounded on all sides by an exceedingly rare and elastic body, called the atmosphere, extending with a diminishing density to an unknown distance into space, but pressing upon the earth with a force equal to that of a homogenous atmosphere five and a half miles high.

William Ferrel, Nashville, 1856

1.1   WHAT IS AIR MADE OF?

Air is somewhat mysterious because it is so insubstantial and invisible, and yet vital to our existence. Like everything else in the *known* Universe, air consists of atoms. It should be noted, however, that cosmologists currently think that atoms make up only about 4 % of the Universe; the remainder consists of so-called 'dark energy' and 'dark matter', the natures of which are currently unknown. The most elementary model of an atom comprises a central nucleus which carries a positive electrical charge, around which various numbers of even smaller negatively charged particles called 'electrons' circle in one or more orbits, like Earth satellites. The number of orbits and the numbers of electrons in each orbit vary from chemical element to chemical element. Most atoms cluster together in groups called 'molecules'. Most gases, such as air, consist of molecules; but there are monatomic exceptions such as helium, with which we fill balloons, and the other so-called 'noble' gases such as argon and neon. Gas molecules are very small; a volume of gas consists almost entirely of empty space. Energy transferred to them by the sun, by combustion and other heat sources, and by being compressed mechanically, causes them to rush around continuously and randomly, in all directions. [Note: In theoretical physics, quantum theory, introduced in the 1920s, has displaced the concept of electrons and other atomic particles as being discrete elements of matter; but, for the purposes of this book, we may assume the elementary model].

Uncontaminated air is made up mainly of two types of gas. At sea level, roughly 78% is 'nitrogen' and 21% is 'oxygen'; the remaining 1% is made up of minute traces of other gases, including carbon dioxide. Atmospheric air also contains various concentrations of pollutant gases such as nitrous oxide from engine exhausts, methane from agricultural and mining activity and chlorofluorocarbons (CFCs) used in chemical processes and refrigeration systems. The air also contains particles of various kinds, such as earth dust and smoke, and various concentrations of water molecules in the form of vapour, or as a suspension of minute droplets, such as clouds, mist and fog.

The diameter of an air molecule is such that three million placed side by side in a row would cover a length of just over one millimetre. In a volume of sea-level air the size of a sugar lump, there are about a thirty million, million, million molecules. Mathematicians, engineers and scientists construct models (simplified, idealised representations) of materials so that they can analyse and understand their behaviours. It is clearly quite impossible to describe precisely the motion of each molecule in, for example, the airflow over a new design of aircraft. But it is necessary to model the flow to predict the aerodynamic quality of the design. Fortunately, it is scientifically valid to employ an alternative, non-molecular, model of the air by which it is represented as a *continuum*; this is a continuous, voidless medium comprising 'particles'. Particles are imaginary entities whose states, properties and behaviours represent the averages of those of the horde of molecules in a small region surrounding the assumed position of a particle. This model will feature large in Chapter 3 on 'Aerodynamics'.

Some people think that air is full of carbon dioxide; but air contains only a tiny bit of this gas - on average, about four parts in ten thousand by volume. This is fortunate, because excessive concentrations of carbon dioxide gas in the atmosphere would displace the lighter element oxygen and asphyxiate oxygen-dependent creatures such as ourselves. In fact, an escape of carbon dioxide gas from the bed of Lake Nyos in Cameroon, West Africa, in 1986, formed a cloud several tens of metres deep that killed over 1500 local inhabitants in their sleep, and almost all other creatures within a radius of 25 km.

1.2   WHERE DID AIR COME FROM?

The layer of gas surrounding the earth is called the 'atmosphere'. The molecules of the atmosphere are prevented from rushing off into outer space by gravitational attraction that pulls on then with a force called 'weight' - as it does on you and everything else on Earth. The Moon has no atmosphere because it is much smaller than the Earth and its gravity is six times weaker than that of Earth, as you could see when the American astronauts jumped around instead of walking. Any atmosphere that the moon originally had escaped soon after its formation.

It is believed, but not definitely known, that the Earth has had three distinctly different atmospheres as it cooled from its original state as a lump of molten rock. The first one probably consisted principally of the two lightest gases, hydrogen and helium. The speeds of most of their molecules were high enough to allow them to escape from Earth's gravitational field, just like rockets that are sent into outer space.

The second atmosphere probably contained steam, carbon dioxide, sulphur and ammonia that escaped from volcanoes, together with some residual hydrogen. They created a greenhouse effect that prevented a large proportion of the heat energy of the Earth's surface caused by solar radiation from escaping back into space, and kept the planet at a temperature of about 70° Celsius – like hot soup.

The modern atmosphere largely owes its makeup to the emergence about 2.7 billion years ago of minute, single-cell, creatures (microbes) called 'cyanobacteria'. In the 'Great Oxidation Event', they used the energy from the sun in a process called 'photosynthesis' by which they acted on the carbon dioxide and water to release oxygen (O_2) and hydrogen (H_2). It had been thought in the past that the hydrogen transformed the carbon dioxide into organic matter, causing carbon dioxide levels to plummet. However, recent research suggests that much of the hydrogen released during photosynthesis escaped into space instead of forming oxidisable hydrocarbons. Recent fossil discoveries in the United States, at Gilboa, have revealed the dramatic effect that the development and rapid spread of photosynthesising, fern-like plants called Wattieza trees and their more successful contemporaries called Archaeopteris, that lived around 380 million years ago, had on decreasing the proportion of carbon dioxide and increasing the proportion of oxygen in the atmosphere. The effect of this 'greening of the Earth' is briefly described in the following chapter.

Ultraviolet radiation from the sun acted upon oxygen high in the atmosphere to produce ozone (O_3) that shields the Earth's surface from the harmful effects of this potentially damaging radiation. This is why there was great concern about development of the hole in the ozone layer that was largely caused by CFC gases used in aerosols and refrigeration systems. The hole is now beginning to close up. The protection offered by the ozone layer allowed the more efficient multi-cellular, photosynthetic, green plants with which we are familiar to cover the earth's surface and to provide much of the oxygen that we breathe today. And, of course, it subsequently allowed animals to evolve. Microbes called 'phytoplankton' that exist near the surface of the oceans convert carbon dioxide and water into carbohydrates and oxygen and are still so abundant that they generate about half the oxygen in the air that we breathe – we damage them at our peril. It is believed that the build-up of oxygen in the atmosphere was hastened by the burial and submersion in water of much of the organic material that would otherwise have been oxidised. This is the origin of the vast coal deposits of the world.

1.3   GLOBAL WARMING

The Earth is maintained at a global average temperature of about 15°C by the 'greenhouse effect' whereby heat energy absorbed by the surface of the Earth from the Sun is inhibited from re-radiating away into space by various components of the atmosphere, of which the most effective is water vapour. It is hypothesised, but not firmly established, that at various times between about 850 and 639 million years ago the atmospheric composition was such that the greenhouse effect did not operate to any significant degree and the ground surface temperature ranged between zero to 10°C in the equatorial belt and averaged about –80°C in the polar regions. The whole global surface would have been covered by ice. Much of life on Earth as we know it could not have evolved or even survived during this period. We are here today because the 'big freeze' was followed by a 'big thaw' which was almost certainly probably driven by volcanic activity. More recently – about 300 million years ago – the roots of primitive trees burrowed into rocks and dissolved them with acids. The material was slowly washed into the oceans where it combined with dissolved carbon dioxide. The concentration of CO_2 in the atmosphere dropped from about twenty times to about twice its present value which reduction suppressed the greenhouse effect, resulting in another major ice age when the ice approached the tropical regions. The process was self-limiting since it weakened as the temperature fell.

Approximately 250 million years ago, life on Earth suffered an enormous setback over a period of about a million years in what is known as the end-Permian extinction. Extreme volcanic activity, known as the 'Siberian straps', pumped enormous amounts of CO_2, together with toxic gases such as fluorine and chlorine, into the atmosphere. This, together with the release of methane from methane hydrate deposits on sea beds, raised the temperature of the ocean and the atmosphere to such a level that mass extinction of the majority of species then living were wiped out. About 95% of marine species perished; more than two thirds of terrestrial reptiles and amphibians died; most insect species were destroyed and a large percentage of plants succumbed. Clearly, some life clung on, including quite a few species of fish and fungi. It is possible that the impact of a large meteorite made matters worse, although it seems unlikely to have been the principal cause.

Today, the greenhouse effect is being strengthened at an ever increasing rate by changes in the atmosphere that are attributed largely to human activity. This phenomenon is known as 'Global Warming'. In addition to water vapour, the principal 'greenhouse gases', as they are called, are methane (CH_4), nitrous oxide (N_2O), the chloro-fluorocarbons (CFCs) and carbon dioxide (CO_2). Apart from water vapour, CO_2 currently makes the largest contribution to global warming, but methane is about twenty times more potent in this respect and is accumulating in the atmosphere at twice the rate of carbon dioxide. Much of the carbon contained in the second atmosphere was locked into underground oil, coal and gas deposits: we are releasing it again in the form of carbon dioxide by the process of combustion of hydrocarbon fuels. Methane is released by burning vegetation for land clearance, by gas and oil exploration activity, by landfill, by wetlands and by the flatus of farmed animals. The latter contributes about 18% of greenhouse gas emissions which exceeds that from all forms of transport. Of greatest concern is that vast amounts methane will be released into the atmosphere if - and almost certainly when - the enormous areas of the Earth's permafrost are progressively melted by global warming. The remains of plant materials currently locked up in the permafrost will be freed to ferment in the depths of the resulting lakes to form methane gas. Methane plumes have recently been observed to emerge from arctic seabed as the overlying ice melts. The process will be self-accelerating as the methane produces increasingly rapid global warming. The effects on the atmosphere, and the implications for the future of life on Earth, are unknown. Most of the nitrogen in the atmosphere came from the sun's action on ammonia released by volcanoes. Nitrous oxide is generated by fossil fuel combustion, mainly by power stations, by the use of nitrogenous fertilizers and by the decomposition of animal waste. The use of CFCs has been limited by international agreement, but the release from discarded refrigerators, particularly in Africa and some parts of Asia, is exacerbating the problem.

1.4   HOW HIGH IS THE ATMOSPHERE?

As height increases, the atmosphere gets thinner (less dense) because the molecules are, on average, further apart. The air pressure also falls; this is because a layer of air at any height has to support the weight of all the air above it. (Imagine being part in a human column in a circus. The strongest man is placed at the bottom, and the lightest at the top.). The variation of air pressure with height as a fraction of the pressure at sea level is shown in the following table.

fraction	(m)	(ft)
1	0	0
1/2	5,486	18,000
1/3	8,376	27,480
1/10	16,132	52,926
1/100	30,901	101,381
1/1000	48,467	159,013
1/10000	69,464	227,899
1/100000	96,282	283,076

The average air pressure at sea level is about 100,000 newtons per square metre (100 kilopascals), also known as one 'atmosphere' or one 'bar', which is nearly one ton per square foot. Some very thin atmosphere exists even above 1000 kilometres (km), which is about one hundred times the height at which long-range airliners fly. There it causes Earth satellites to lose energy very slowly through aerodynamic drag. About 70% of the mass of the atmosphere lies below 10 km, which is a bit higher than Mount Everest, and 99.99999% lies below 100 km.

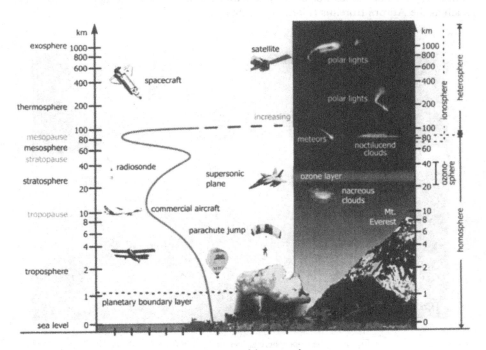

Fig. 1.1 Layers of the atmosphere

Atmospheric layers are illustrated in Fig. 1.1 and defined as follows:

Troposphere: from the Greek for 'mixing'. This extends from the surface to between 7 and 17 km. The temperature generally decreases with height. The air in this layer is subject to mixing due to solar heating of the Earth's surface. Hot air rises and, as it cools, the latent heat of the water that it contains is released to lift the air further, until all the water has been removed.

Stratosphere. This extends from the upper boundary of the troposphere to about 50 km: the temperature increases with height.

Mesosphere. This extends from about 50 km to about 85 km: temperature decreases with height.

Thermosphere. This extends from about 85 km to over 650 km: temperature increases with height.

Exosphere: This extends from about 700 km to 10,000 km: here, temperature loses conventional meaning because the air molecules are, on average, so far apart.

This book concentrates on the troposphere where most of the phenomena and processes that we on the ground can observe occur. Its thickness relative to the radius of the earth is equivalent to a 0.4 mm thick film covering a soccer ball. The outer faint layer between the dark sky and the clouds in Fig. 1.2 gives an idea of its thinness. The book does not cover phenomena that occur in the upper atmosphere such as the Aurora Borealis (Northern Lights).

Fig. 1.2 The troposphere is seen in the thin faint outer layer

1.5   HOW DO SOLIDS, LIQUIDS AND GASES DIFFER?

It should first be said that almost all substances *can* exist in solid, liquid and gaseous forms, given the appropriate conditions of pressure and temperature. There exists a fourth state of matter called 'plasma' which is produced by the process of 'ionisation' in which normal atoms are either stripped of one or more electrons, to produce 'cations', or forced to accept more electrons to produce 'anions'. Gases may be ionised by heating them to extremely high temperature. Liquids may be ionised by applying a large voltage difference across them. Ionisation of gases and liquids increase their capacity to conduct electricity; a common example is the phenomenon of lightning. We shall meet plasma again in Chapter 6.

When two normal gas atoms are very far apart, there is negligible interactive force between them. As the separation distance decreases, an increasingly strong attractive force operates. If this were the only force, nothing but extremely dense masses of super solids would exist. [If all the empty space between the molecules that constitute the entire human population were removed, we would all fit into one matchbox.] However, as the separation distance between atoms gets smaller and smaller, an increasingly strong interatomic repulsive force comes into action. So, at some separation distance, the two forces cancel out and this is the so-called 'equilibrium distance'. Atoms tend to cluster at this separation distance where they are said to be 'bound'. A moving mass possesses energy of motion, called 'kinetic energy'. The amount of energy that has to be transferred to a pair of bound atoms to separate them is called the 'dissociation' energy. This varies with the type of atom. It is proportional to the product of the mass and the *square* of its speed. Free (unbound) atoms possessing more than the dissociation kinetic energy that impact the cluster can free individual atoms.

The speeds of motion of gas molecules are distributed over a range of values. The temperature of a volume of gas is proportional to the average of the square of those speeds. An increase in temperature corresponds to an extension of the range of speeds to higher values with a higher proportion of the molecular speeds lying in the upper part of the range. If the temperature of a substance is sufficiently high that the kinetic energy of free molecules greatly exceeds the dissociation energy of the material, it is not possible for permanent clusters of molecules to form. This is the *gaseous* state. Where the energies of the free molecules are evenly distributed about the dissociation energy, clusters will be formed and last for some time, but eventually will be disrupted and new clusters will form. This is the *liquid* state. In the *solid* state, the dissociation energy is not exceeded and all molecules are bound into a single cluster. The molecules of most solid materials lie close together and are held in position by strong forces acting between them. Unlike fluids, solids resist changes of shape and being split. Solids resist forces acting so as to deform them and many, but not all, revert almost to their original form when deforming forces are removed. This property is called 'elasticity'. The molecules of solids can, however, vibrate. The strength of vibration increases with temperature which can be raised by the addition of energy in the form of heat or by means of work being done on them, as can experienced by your fingers in rapidly bending and unbending a piece of soft wire. If solids are supplied with enough heat, the molecules break free of their bonds and the solid melts into a liquid. As you know, they vary in their ability to do so.

Gases and liquids are both categorised as 'fluids' because they flow, unlike solids. The bonds between the molecules of liquids are continuously formed and broken, which allows them to flow and shear and easily deform so that they take up the shape of a container. However, the attractive forces between liquid molecules are sufficient to prevent the majority of molecules escaping to form a gas and to completely fill any container into which they are put. If a liquid is heated sufficiently, the molecules acquire so much energy that they break free from their forces of mutual attraction to become vapour and, eventually, gas. Vapour is defined as a gas at lower than its 'critical temperature' above which it is impossible to liquefy it purely by increase of pressure. This process is clearly seen when a pan of water is heated to produce wet steam. Further heating of the steam will eventually produce dry, super heated steam.

Gases consist of free atoms or molecules and can easily be moved around - for example, by your lungs. The molecules of gases do not attract each other strongly, except when in very close proximity, and they rush around all over the place, frequently bumping into others and bouncing off them, for the reason explained above. Because they are free to rush around, they completely fill any empty container that you put them into, unlike the water in a half full (or half empty) container. This is why gases flow everywhere that they are free to go.

1.6   HOW DO WE DESCRIBE THE STATE OF A GAS?

The state of a gas is described and measured in terms of a number of physical quantities, of which the most common are *density*, *temperature* and *pressure*. We shall discuss the representation of the state of motion of a gas in Chapter 3. Density is defined as the mass per unit volume (kilograms per cubic metre). In any pure material, it is equal to the number of molecules per unit volume multiplied by the mass of each molecule. As we have seen, air is a mixture of nitrogen and oxygen molecules. Each molecule comprises two atoms, as indicated by their chemical formulas N_2 and O_2. The relative molecular masses and relative proportions of these gases in air at sea level produce a density of about 1.2 kg/m^3. In comparison, the density of water is about 1000 kg/m^3 and of steel is about 7800 kg/m^3.

'Mass' is difficult to explain in everyday language. We can get a feel for meaning of mass if we consider the relative difficulties of throwing or kicking a tennis ball and a football. It is a measure of the difficulty of changing the speed or direction of travel of an object. Its unit is the 'kilogram' (kg). It is most important not to confuse 'mass' with 'weight', which is a measure of the force applied to a mass by the gravitational field of the Earth. Its unit is the 'newton (N)' which is the weight of an average sized apple. At the Earth's surface, the weight of a body in newtons is 9.81 times its mass in kg. (The common commercial practice of indicating weight in kilograms is scientifically incorrect; but it will only become a practical problem in the far distant future for shoppers on another planet). Out in space, far from any large celestial objects, the weight of an astronaut is negligible; but his or her mass is unchanged and it's just as difficult to change his or her speed or direction of travel as it is on Earth. Astronauts in Earth orbit are not 'weightless', because the gravitational force keeps them in orbit; without it they would simply fly off at a tangent in a straight line. They are in 'free fall' and appear to be weightless because they feel no reaction forces from the surfaces on which they 'stand' or 'sit'. Chapter 3 explains that a mass will only follow a curved path if a force is applied to it at right angles to the direction of travel, as with a stone whirled round on the end of a string. Similarly, artificial gravity may be produced by rotating the spacecraft at an appropriate speed forcing the astronauts to 'stick' to the wall.

As previously mentioned, the temperature of a gas is proportional to the average kinetic energy of random motion possessed by *individual* molecules. Temperature is quite distinct from heat, although in everyday speech we use the word 'hot', to indicate high temperature. In strict scientific terms, heat is the *process of transfer of energy* from one system to another connected system which occurs when the temperature of the first exceeds that of the second.

Mechanical work is done on a system by a mechanical force applied to it when the point of the system at which the force is applied moves in the direction of the force. You do work on the air in a bicycle pump when you push the plunger in against the air pressure. This produces rises in air pressure, density and temperature: you can feel the last mentioned. In the 1843, a Manchester brewer named James Joule demonstrated that work and heat are equivalent means of transferring energy by employing the fall of weights to rotate a paddle in a container of water and relating the work done by gravitational force on the weights to the temperature rise of the water. The same temperature rise could have equally been produced by heating the water by means of a gas flame: thus the 'equivalence'. We are happy to think of doing work as a *process* of transferring energy from the 'worker' to the 'worked', and not to think of work as energy that flows between the two. Similarly, 'heating' some material is a *process* of energy transfer. However, it is difficult to consider the concept of work's physically equivalent quantity, heat, as a process, because the concept of heat as a *form* of energy that is possessed by systems and flows between systems is so deeply embedded in our everyday speech that it is difficult to avoid its use - and it will not always be avoided in this chapter. Thermodynamicists look away now!

The possession of energy by a substance or body gives it the capacity to do work. For example, you use chemical energy derived from your food to walk upstairs, because you have to do work against the gravitational force acting downwards on you. In so doing you gain 'potential energy' which is converted into kinetic energy by gravitational force acting on your moving body if you fall downstairs. If you stand on one end of a seesaw and someone jumps off a platform onto the other end of the seesaw, his or her kinetic energy, which is generated by their weight doing work on them as they fall, does work on you by sending you up in the air. In the process, you gain potential energy which is progressively converted back to kinetic energy as you fall to earth. An archer expends chemical energy in doing work on his bow to increase its potential energy of bending (just like a compressed spring), most of which is converted into kinetic energy of the arrow when it is released. The arrow can do work on a victim with rather unpleasant results – but where does the kinetic energy of the arrow go? It goes partly into the kinetic energy of the resulting backward motion of the victim and partly into generating heat in the victim's body as it distorts and ruptures his tissues. These examples illustrate the 'Principle of Conservation of Energy' which expresses the fact that energy cannot just disappear - it can only be converted from one form into another – and ultimately into heat. (Here we exclude nuclear reactions such as the atom bomb in which a small mass is converted into a vast amount of energy according to Einstein's famous equation $E = Mc^2$, where c is the speed of light. This means that one gram of any substance could, in principle, be converted into 90,000,000,000,000 joules which approximates the energy in 700,000 gallons of gasoline (petrol)).

Interstellar gas, in which molecules are few and far between, can have a very high temperature due to kinetic energy gained by the molecules from stellar radiation, but their energy *density* is very low. Hot things transfer heat to cooler things by transferring molecular energy through interaction between the molecules of the two materials. It is the increased molecular movement that excites the nerves of the skin and causes the tissue damage and pain of a burn. A familiar (non-gas) example that illustrates the difference between temperature and heat is provided by the firework sparkler: the temperature of each small spark is very high, but it does not painfully burn the skin because the total molecular energy transferred to the skin by each spark is very small. Another is the fact that the heat energy of a warm bath is many times that of a burning match, but the temperatures are very different, as you well know. Temperature is measured according to a number of different scales of which the most familiar is the Celsius scale used by meteorologists. However, physicists also use another scale that measures 'absolute' temperature in Kelvin. This scale has a zero at –273 Celsius (^{o}C) at which temperature all molecular movement ceases. So a temperature t on the Celsius scale equals $(273+t)$ Kelvin (K) on the scale of absolute temperature. At zero K, all atomic activity would cease.

The term ' air pressure' is familiar from meteorology reports and weather forecasts because the weather is strongly related to the spatial variations atmospheric pressure, as well as of temperature. But what does air pressure mean and what is its origin? We have seen that gas molecules rush about randomly in all directions, typically at speeds of hundreds of metres per second. Air molecules gather most of their energy from the sun. As previously stated, gas molecules do not actually collide because, as they get very close to each other, strong repulsive forces are generated. Such close encounters alter the speed and direction of travel of each of a 'colliding' pairs of molecules. A good analogy is provided by balls on a snooker table. Imagine that there are hundreds of balls on a snooker table and a few players distributed around the table keep hitting them. Each cushion will be hit by a succession of balls and these create a succession of forces on the cushion. In the same way, any solid surface exposed to the air is battered by millions and millions of molecules which are repelled by the surface molecules of the solid: in 'bouncing off' the surface they generate a virtually continuous force on it. The pressure of air is a measure of the amount of force per unit area that its molecules impose on any surface exposed to it: its units are 'newtons per square metre', also known as 'pascals', after a French scientist. As we shall see later, an aircraft is held in the air by a difference of average pressures acting on the upper and lower surfaces of the wings. Just think, you are kept aloft by the impacts of millions and millions of minute molecules on the wings. Pressure is not confined to the interface between a fluid and a contiguous solid surface. Pressure is exerted by one region of fluid on a contiguous region by virtue of the forces of repulsion exerted by the molecules of one region in interacting with those of the other.

You experience air pressure when you blow out your cheeks. You are increasing the density of the air (the number of molecules per unit volume) to the point where the pressure produced by muscular contraction of the chest is balanced by the tension in the cheeks that sustain that pressure. When you fly in an airliner, at a height of about 10,000 metres, the air pressure in the cabin is artificially increased to equal that that a height of about 3000 metres so that you take in a sufficient number of oxygen molecules to allow you breathe comfortably. The air pressure at a typical flying height is about one fifth of that at sea level and therefore, with no air pressurization, the number of oxygen molecules in each breath would be reduced by a factor of five. The pressure increases as the aircraft lands and this pushes on the eardrum which lies at the inner end of the ear canal. This force is normally balanced by air pressure transmitted along the Eustachian tubes that connect the oral cavity with the other side of the eardrum. If these are blocked, the unbalanced force on the eardrum can cause severe distress. They can usually be unblocked by holding one's nose, closing one's mouth, and increasing the air pressure in the chest.

Pressure exists everywhere in a gas (and in any fluid). In a gas it is proportional to the total kinetic energy of random motion of all the molecules *in unit volume*; this is termed the kinetic energy *density*. As we have seen, gas temperature, is proportional to the average kinetic energy *per molecule*. Hence, the pressure of a gas at a given temperature is proportional to the number of molecules in unit volume; that is, to the density of the gas. These relations are the basis of the so-called 'Universal gas law' that applies to all gases under almost all conditions. As previously mentioned, air at any height has to support the weight of the air above it. This is why the density, and therefore pressure, of atmospheric air at a given temperature, decrease with height. In terms of molecular motion, this pressure gradient is explained by the fact that molecules existing in any horizontal slice of the atmosphere, and moving upwards, lose kinetic energy because gravity does negative work on them, just as it does on a ball thrown into the air; and molecules travelling downwards similarly gain kinetic energy. Hence, the molecular kinetic energy density, and therefore the pressure decreases with height. Water (hydrostatic) pressure increases with depth for the same reason.

The average atmospheric pressure at sea level is 100,000 newtons per square metre (nearly one ton per square foot or about half of the air pressure in a typical car tyre.). The surface area of a typical adult Caucasian man is about 1.8 square metres (19 square feet), so the total compressional force on the body surface is nearly 20 tons. So, how is it we don't get crushed? Well, fortunately, most of the tissues and fluids of the body are, on average, also at atmospheric pressure, so the external force is balanced. When a diver descends into water the pressure on and within the body increases, one effect of which is that nitrogen gas in any inhaled air gets dissolved in the blood. If the diver returns too quickly to the surface the nitrogen comes out of solution, forms bubbles in the blood and causes the 'bends', which is life threatening.

Air can be liquefied by putting it under high pressure and cooling it to below −216 °C (57 K). The average spacing between the molecules, together with their kinetic energies, are so greatly reduced that mutually attractive forces are sufficient to allow them to cluster together, at least temporarily, as is characteristic of liquids. Liquid air is a little less dense than water at 870 kg/m^3 (compared with about 1.2 kg/m^3 in its gaseous form at sea level). It is used for cooling other substances and as a source of nitrogen, oxygen and argon.

1.7   HOW DOES HEAT TRANSPORT ENERGY IN AIR?

Heat transfers energy in air by means of two mechanisms. The molecules of hotter (higher temperature) regions of air on average possess more kinetic energy than the molecules of colder regions. When there is a temperature difference between neighbouring regions of air, collisions between their molecules at the interface increase the energies of those in the cooler region and decrease the energies of those in the warmer region, and so even out the temperature differences. This process of energy transfer is called 'heat conduction'. The temperature of a region of *still* air in which conduction dominates the process of energy transport varies smoothly from place to place. One can subjectively sense this form of energy flow process, also called 'diffusion', if one runs a tepid bath and then increases the temperature of a slowly running hot tap. One can feel the 'heat' being transported away from the falling water and moving up one's body. Air does not conduct 'heat' efficiently because its molecules are, in microscopic terms, on average so far apart and so rarely collide. This is exploited by double glazing, by birds when they puff up their feathers in winter, by woollen sweaters, by hollow animal hairs used in jackets, by eiderdowns that originally contained the feathers of eider ducks and by duvets, which all contain and retain considerable amounts of air.

The other mechanism of 'heat' transport, termed 'convection', involves the bulk movement (flow) of air in which locally hotter regions containing billions of molecules move into the proximity of, and mix with, cooler regions. The fundamental process of energy exchange is still due to molecular interaction, but the bulk mixing motions are much more effective than conductive diffusion. Convection may either free or forced. Free atmospheric convection plays a very important role in influencing the weather and is partly responsible for winds, cyclones and thunderstorms as well as very large-scale global movements of air masses, as described in Chapter 5. It is caused by uneven solar heating of the ground surface which, in turn, heats the local air. At a given pressure, a higher temperature implies a lower concentration of molecules (density). These molecules are displaced by the more densely distributed molecules of cooler surrounding air and rise. This mechanism causes large scale mixing of air masses of different temperatures. The rising columns of warm air that are called 'thermals' assist glider pilots and many types of large bird to circle and climb effortlessly. As explained in Chapter 3, turbulence is a chaotic motion of fluids which involves intensive mixing of fluid elements that possess different speeds. Atmospheric turbulence plays a major role in reducing atmospheric temperature gradients and also disperses pollutants.

Forced convection is generated by artificial means of moving air, such as fans, so that hotter and cooler volumes of air mix in a turbulent fashion and exchange energy. The promotion of convective 'heat' flow by turbulence is exploited in industrial and domestic 'heat' exchangers. An example of forced convection in water is that employed by a bather to cool down water from an over-hot tap flow by means of paddling the water with the hands. It is obvious why the overheated bather does not wait for diffusion to relieve the discomfort.

Solar radiation is an electromagnetic wave phenomenon that, like light waves or radio waves, does not depend upon support by a material medium, unlike sound (see Chapter 4). It can transmit energy to the molecules of material media as we know to our cost if we spend too long in the sun without adequate protection. Thermal radiation of energy by a hot body increases rapidly with increase of absolute temperature. So-called 'radiators', filled with hot water or oil, that are used to heat buildings, operate principally by free air convection, although they do radiate a little of their 'heat', as you can feel when close to them. The much hotter elements of old-fashioned electric 'fires' do principally radiate heat because the elements run at a much higher temperature.

1.8   WHAT ARE THE ELECTRICAL PROPERTIES OF AIR?

Dry air is a very poor conductor of electricity, fortunately for us. However, once the voltage of an applied electric field is increased to the 'breakdown value', free electrons, that have a negative electric charge, become sufficiently accelerated by the electric field to create additional free electrons by colliding with, and ionizing, neutral gas atoms or molecules in a process called 'avalanche breakdown'. The breakdown process forms a plasma that contains a significant number of mobile electrons and positive ions (atoms with electrons missing), causing it to behave as an electrical conductor. In the process, it forms a light-emitting conductive path, such as a spark, arc or lightning.

Air in a natural state contains water, either in gaseous form or in the form of a suspension of minute droplets. There is a limit to the concentration of water molecules that air at any particular temperature can hold in gaseous form; this limit increases with temperature. The ratio of the actual concentration to this limit (saturation) is called 'relative humidity'. In Britain, the relative humidity typically ranges from about 70% on a dry sunny day to 100% on a misty, wet day. If the limit is exceeded, the air molecules are not sufficiently energetic or numerous to stop (dissociate) the water molecules from gathering together (condensing) to form clouds, mist, fog and condensation on a cold window. We describe the white mist that forms near the spout of a kettle of boiling water as 'steam': technically, it is a vapour not a gas. Only the invisible steam that emerges from the spout before it mixes with the air, cools, and forms the mist, is truly a gas. [Have a look at it.] The electrical conductivity of moist air increases rapidly with increase of relative humidity. The effects can sometimes be seen and heard in the vicinity of pylons carrying high voltage cables

1.9   WHY IS THE SKY BLUE?

We don't usually think about it, but the Earth would be a very different place if air were not almost transparent. You might like to speculate about what types of creature might have evolved if that were the case. But why is dry air so transparent? This is a difficult question to answer because the nature of light is itself difficult to understand. In simplistic terms, it is because, in the absence of substantial suspensions of water molecules and droplets (as in mist and fog), the elementary packets of light, called 'photons', are not very strongly scattered (redirected) or absorbed (lose energy to the electrons of the medium) by the air molecules during passage over distances that are important for everyday human vision. This is partly because the air molecules are, on average, far apart compared with the range of wavelengths of visible light. The electrons surrounding an isolated atomic nucleus oscillate at frequencies that are characteristic of that element. The incidence of light energy (protons) on the atom can set the electrons into oscillation which will be strong if the frequency range of the light encompasses one or more of the natural oscillation frequencies. In this case, the energy of incident protons is taken up by the electrons and is subsequently either dissipated (absorbed) as heat or re-radiated (scattered) The frequency range of visible light does not include the characteristic frequencies of the electrons of oxygen and nitrogen atoms; hence the transparency. Moist air is less transparent because of the interaction of light with water molecules is somewhat stronger than those of air. We are made aware of some degree of weak scattering over long distances by the tonal recession towards the blue end of the spectrum exhibited by the distant landscape, as depicted in paintings.

The atmosphere appears to be transparent judging by views of the Earth from space in which the landmasses, oceans and clouds are clearly seen. But if the air really were completely transparent, the sky would appear to observers on Earth always to be black, as on the Moon that has no atmosphere and no water vapour or clouds. On clear sunny days, the sky appears blue; and it often takes on various yellowish/reddish tones during and after sunset. The reason was identified by the Irish physicist, John Tyndall, in 1859, and subsequently explained by mathematical analysis by Lord Rayleigh in 1873. Light from the sun has to pass through many kilometres of atmosphere. During this passage, air molecules scatter the light into all directions, most strongly in the lower atmosphere because air density falls with height.

The degree of scattering of waves, such as those of light, sound and water, depends strongly on the ratio of size of the scattering object to the wavelength of the wave. A familiar example of wave scattering is seen in a fairly quiet sea near to the bases of cliffs where individual lumps of rock protrude through the water surface. Approaching waves pass almost unscathed around *small* rocks, with just a hint of weakly scattered waves spreading out in all directions from the obstacle. In the case of *rather larger* rocks the scattering process is rather different. Waves passing the edges of the rocks 'bend' around the edge and appear to originate from the edges. This phenomenon is scientifically known as 'diffraction'. Diffraction of sound enables neighbours to converse over a garden wall even if it is too high for them to see each other. Diffraction also operates in the case where a wave is incident upon a small gap between fairly large rocks; much of the wave is reflected, but some passes through and spreads out from the gap in a circular pattern. Rocks that present obstacles that are *very large* compared with the wavelengths of the approaching waves strongly reflect them – that is, they send most of the wave energy back where it came from. Blue light has the shortest wavelength in the visible spectrum, apart from violet and indigo. Scattering of the blue light is most apparent because the eye is more sensitive to it than to violet and indigo. Sun light falls on all parts of the canopy formed by the sky and light is scattered in all directions from all parts. Thus, as well as seeing sunlight directly (which on clear days we do not do for obvious reasons) we see sunlight indirectly via scattering, and a clear sky in the daytime appears blue everywhere: the tone lightens towards the horizon (check it out –any idea why?).

The atmosphere carries dust, water and ice particles in addition to the much smaller air molecules. As the sun sets, the light travels an increasingly long distance though the atmosphere to our eyes. Transmission of the shorter wavelengths is suppressed, leaving the longer scattered wavelengths (yellow, red) to make the journey to the surface of the earth. This phenomenon is most obvious once the sun has set beneath the horizon; for some time, the sky remains lit by light scattered from the clear air and from clouds, which eventually fades into darkness. In the absence of light scattering and clouds complete darkness would fall as soon as the sun set below the horizon. On overcast days, the sun's light passes through suspensions of water droplets in the clouds that are considerably larger than the wavelengths of the all the colours of the visible spectrum, which are therefore all scattered to give the familiar grey appearance.

Rayleigh scattering effect produces the blue of the sky. Yet the sunlight reflected by the Earth's surface, and seen by astronauts in orbit or in deeper space, does not appear blue, except for the water in the oceans. Pure water actually has a naturally blue tinge, which is obvious only when great depths of water are seen from above. The light reflected into space by terrestrial features has passed twice through the atmosphere. So why are the landmasses and clouds seen by in their normal colours.? [Note: many images of the Earth as seen from space have been manipulated to produce 'false colour' pictures in order to highlight particular features]. It turns out that molecular scattering is extremely weak and any effect of the double passage on the appearance of terrestrial features seen through the atmosphere is negligible. However, one-way scattering is sufficient to replace the blackness of space by a beautiful blue canopy for our illumination and delectation on Earth.

It might be thought that rainbows are nothing to do with air *per se*. However, they are caused by the fact that light travels faster in air than in water. This causes the phenomenon of refraction which bends light rays as they pass across an interface between the two media. Refraction causes a stick pushed through the surface of water to appear bent and makes the depth of a swimming pool seem smaller than it actually is. The principal arcs of rainbows are caused by double refraction and internal reflection of sunlight by raindrops. As light rays from the sun behind the observer enter the 'front' of raindrops in the view of the observer, their paths are bent towards the centre of the drops. They then reflect from the 'back' of raindrop and are again refracted by a similar amount as they leave the 'front' to travel to the eye of the observer. As Newton demonstrated, light waves of different wavelengths are refracted to different degrees. So the eye sees the various colours of the spectrum at slightly different angles, producing the familiar spread of the rainbow. However, if the average diameter of the water droplets is less than about 100 micrometres, they diffuse the light and form diffraction bands that differ from wavelength to wavelength. These overlap and the cumulative effect is to disrupt the refractive effect to such an extent that a white 'fogbow' is formed [see www.atoptics.co.uk/droplets/fogbow.htm].

2

Air: the Supporter of Life

Wild air, world mothering air,
Nestling me everywhere.

Gerard Manley Hopkins 1879

2.1   OXYGEN, THE GIVER OF LIFE

Although not all life forms on Earth depend upon oxygen (it can actually be toxic to some), it is essential to the existence of the vast majority of complex, multi-celled organisms. Most land animals absorb oxygen from the air. Some marine inhabitants, such as whales and seals, also breathe in atmospheric air when they break surface. In the absence of extreme atmospheric pollution by volcanic events or industrial operations, the oxygen content of the air varies little. Most marine animals absorb oxygen from the water in which they are immersed. This makes sea creatures very vulnerable to chemical pollution and to climatic phenomena such as ocean warming that reduce the oxygen content of the water.

The proportion of oxygen in the atmosphere has varied widely over the life of the Earth - about 4.6 billion years. As explained in the previous chapter, primitive life forms were operating photosynthesis as long ago as 2.7 billion years. But it was the evolution of trees with photosynthesising leaves around 375 million years ago that dramatically increased the amount of oxygen in the atmosphere, up to a proportion of about 35 percent. In so doing, it stimulated the evolution of very large insects, such as dragonflies with wingspans of one metre, two-metre long millipede-like creatures, and other lungless invertebrates. It is hypothesised that it was the great increase in oxygen generated by the spread of trees across most of the Earth's surface that allowed some of the vertebrates, all of which at that time lived in the sea, to 'test the land and the air'; and for some to find them to their liking. Fortunate for us!

2.2   WHAT HAPPENS WHEN WE BREATHE?

When we expand our chest cavity, the air pressure in the thoracic cavity between the lungs and the chest wall falls. The higher atmospheric pressure of the external air pushes it into the lungs. It travels through the branched bronchial tubes of progressively decreasing diameter and finally into the 300 million alveoli, which look like clusters of little balloons and have a very large surface area (about 100 square metres in male adults). Each alveolus is wrapped in a fine mesh of blood-filled capilliaries covering about 70% of its area. The alveoli have radii of about 0.05 mm which increases to around 0.1 mm during inhalation. They are particularly vulnerable to irreversible damage by tobacco smoke. The alveoli are inflated and air exchanges oxygen molecules with the haemoglobin in the red blood cells in the small pulmonary capillaries that run between the alveoli and are separated from them by an extremely thin membrane. Carbon dioxide passes in the opposite direction to be exhaled. Shortly after inhalation finishes, the chest muscles relax and exhalation occurs automatically. The red blood cells transport the oxygen to all the cells in the body which rely upon it for their functioning. This mechanism is common to all animals having lungs.

2.3   OTHER ANIMAL BREATHING MECHANISMS

Some animals equipped with lungs also breathe through their skins to some extent. Earthworms and amphibians have skins that are permeable to gases. The latter have simple lungs that need to be supplemented by skin breathing: the proportion of oxygen absorbed through the skin of a bullfrog is about 80%. Boa constrictors absorb about 21% through their skins and even some warm-blooded animals, such as certain bats, absorb up to 13% by means of this mechanism. Some fish, whose primary oxygen collection systems are gills, including plaice and eels, absorb up to 30% of their oxygen uptake directly from the water through their skins. The skin of the unlovely giant aquatic salamander of the U.S. called the Hellbender (also known as the 'Snot Otter') is very wrinkled which gives it a large area furnished with a dense network of subcutaneous capillaries through which it breathes.

All insects breathe though openings in their bodies called 'spiracles' that communicate with networks of channels called 'tracheae' that run throughout the whole body and exchange oxygen with it by means of slow passive mechanisms, including diffusion. However, recent research has revealed that many insects including ants, butterflies and beetles, actively pump air in and out of the tracheae by means of rapid compression and expansion actions of the head and thorax.

2.4   PLANT RESPIRATION

All parts of plants respire through pores distributed over the surface. In leaves these are called 'stomata' and in tree branches they are called 'lenticels'. As explained in the previous chapter, the evolution of strong, searching roots by primitive trees, together with the consequent destruction of rocks and the associated sequestration of carbon dioxide led to a great decline in the atmospheric content of this gas. Remarkably, this aided the development of large, flat plant leaves. Sunlight heats such leaves; their temperature is controlled by the emission of water vapour through their stomata. The density of stomata is controlled by the CO_2 content of the air. The previously very high content militated against such an evolutionary development.

Roots also need oxygen to stay alive. Plants such as rice that thrive in waterlogged soil contain air channels through which it can be channelled down from above. Plant respiration, which continues day and night, involves the absorption of oxygen and the expiration of carbon dioxide.

Air plays a vital role in the process of photosynthesis by green plants in which the energy of sunlight stimulates them to convert water and carbon dioxide into sugar and oxygen, the former being plant food and the latter being released into the atmosphere. Its importance is recognised by the concern about the destruction of tropical forests and the call to plant new forests to counter the excessive build up of carbon dioxide in the atmosphere. Of the carbon absorbed by the trees in a tropical rain forest, only about two and a half per cent is ultimately converted into biomass. About sixty per cent rest is ultimately returned to the atmosphere by leaf, root and wood expiration and about thirty four per cent is returned via respiration by fauna and flora living off the trees. A very small proportion is retained in the earth by an increase of organic matter.

On balance, the amounts of oxygen released and carbon dioxide absorbed during photosynthesis substantially exceed the amounts passing in the reverse direction during respiration, thus contributing crucially to the oxygen in the atmosphere that is vital to most forms of terrestrial life.

2.5   OCEANS AND RIVERS ALSO 'BREATHE'

Atmospheric gases are exchanged with the oceans by means of various mechanisms, of which the principal involves wind-generated waves. Experiments have shown a strong relation between the rate of exchange and the square of the surface slope of the wave components that have very short wavelengths of less than 30 mm. When waves break, air is captured and driven down in the form of bubbles to a depth of up to ten metres. As the bubble rise to the surface some of the gas passes into the water where it is dissolved. This process also occurs on a smaller scale in turbulent river flows. Another exchange mechanism involves near-surface turbulence in the water. The exchange of carbon dioxide across the air-sea interface is a primary influence on the distribution of this gas in the atmosphere. About one third of the atmospheric carbon dioxide produced by human activity since the industrial revolution has been absorbed by the oceans.

A very thorough programme of measurement of the uptake of CO_2 by the North Atlantic, involving 90,000 individual measurements made by specially equipped merchant vessels, has revealed that it halved during the period 1995 to 2005. A similar study of the other principal absorber of CO_2, the Southern Ocean, revealed that it is nearly saturated with the gas and can absorb little more. The cause is not clear. Various hypotheses have been advanced. One widely held view is that the large scale global circulations of water in which colder water sinks and warmer water rises are changing their patterns which affects CO_2 uptake. Some scientists suggest that the amount of carbon dioxide absorbed by the oceans decreases as the water temperature increases, which would exacerbate global warming. However, recent studies of the effect of surface water temperature found little correlation. Others suggest that it is a global increase in salinity that is responsible. Research to resolve the question continues apace.

Every day, the oceans absorb about 30 million tons of CO_2 from the atmosphere; it dissolves in sea water to form carbonic acid in the process termed 'ocean acidification'. The US National Academy of Sciences has recently reported that eight years of measurement off the State of Washington have revealed that acidification of that area of ocean is proceeding at a rate more than ten times that predicted by current climate change models The increasing acidity of the surface layer of the oceans due to CO_2 absorption is not only leading towards saturation and a decrease in CO_2 uptake, but is damaging the ability of the many sea creatures to make chalk-based shells. This has serious implications for the future of crustaceans, such as lobsters, as well as coral reefs and the marine life that depends upon them. Coral reefs provide food resources for 500 million people worldwide. The concentration of oxygen in rivers and seas is of vital importance for marine life. Global warming decreases the oxygen concentration, threatening biodiversity and food stocks.

3

Aerodynamics and Flight

Oh! I have slipped the surly bonds of Earth
And danced the skies on laughter-silvered wings;
Sunward I've climbed, and joined the tumbling mirth
Of sun-split clouds, — and done a hundred things
You have not dreamed of — wheeled and soared and swung
High in the sunlit silence. Hov'ring there,
I've chased the shouting wind along, and flung
My eager craft through footless halls of air. . . .

Up, up the long, delirious burning blue
I've topped the wind-swept heights with easy grace
Where never lark, or ever eagle flew —
And, while with silent, lifting mind I've trod
The high untrespassed sanctity of space,
Put out my hand, and touched the face of God.

John Gillespie Magee, Jr. (American Spitfire pilot)
1941: died aged 19

3.1 THE NATURE OF FLUID FLOW

Gases and liquids are both categorised as 'fluids' because they flow. The defining characteristic of a fluid medium is that the molecules of which it is constituted continuously move from place to place, unlike those in a solid which vibrate about a fixed position and interact strongly to produce resistance to changes of shape. The weaker forces of interaction between the molecules of a fluid medium control their relative motions and they mix with each other allowing the medium to adapt to any geometric form imposed upon it (such as water poured into a glass) and to flow with little resistance, adapting to geometric constraints as it travels (such as water in a piping system or air in a vacuum cleaner). Fluid flow relative to a solid surface is constrained by interaction between the molecules of the fluid and of the solid. This interaction controls the resistance to motion experienced by bodies moving through the air which is expressed in terms of an fluid-dynamic 'drag' force. The work done on the air by vehicle propulsion systems in overcoming this drag is provided by the chemical energy that is supplied in the form of fuel combustion. This energy given to the air is dissipated into the unrecoverable heat energy of random molecular motion. This process costs our pockets, and Planet Earth, dear.

A wonderful compilation of images of fluid flow is cited in 'Suggestions for further reading' at the end of the book.

3.2   WHAT MAKES AIR MOVE?

In order to understand the general physics of fluid motion and to construct mathematical models that allow us to analyse and understand particular cases we need to have a theoretical model that represents the physical properties of pressure and density that were introduced in Chapter One. It is clearly impossible to construct a 'microscopic' model of a fluid which explicitly and precisely represents the motions of each of its component molecules - there are simply far too many. However, it is possible to construct a 'macroscopic' model which is based on the assumption that a fluid is continuous (voidless). For the purpose of representing its kinematic (motional) state a fluid is conceptually divided into minute elements of which the properties and states at any one time represent the average properties and states of the millions of molecules within the boundaries of the element at that instant of time: these elements are called 'particles'. [These average properties and states can be theoretically derived by the use of statistical models, but this need not concern us in this book.] The 'centre of mass' of a particle at any one instant of time is that point at which the system could be balanced on the point of a needle if all the molecules were frozen in their positions at that time

According to Newton's 'Laws of Motion', the motion of the centre of mass of a system containing many interacting particles is influenced only by the *external* forces applied by *outside* agents and not by any interactions between the particles *within* the system. Hence, a fluid element moves as if the total force on it were being applied to a point-like body which is located at the centre of mass and has the same total mass as the molecules within the element. [We exclude considerations of rotational motion and internal vibration of molecules for the sake of simplicity.] This form of model is easier to comprehend when applied to solid systems in which molecular interaction takes place, but molecular mixing does not. Remember that there is no such thing as a perfectly rigid solid material; all solids deform to some extent under the action of applied forces. Consider what happens when you throw a small solid rubber ball. The molecules that form the surface of your throwing hand apply forces to the molecules that form the surface of the ball in contact with the hand. These forces are reacted by the ball molecules which move inwards slightly, thereby imposing forces on the next 'rank' of molecules located just below the surface, and so on, in accordance with Newton's Third Law of Motion: action and reaction are equal and opposite. The resulting distortion of shape passes from molecule to molecule across the ball in the form of a wave. The net result is that the ball moves away from the hand as if it were a single continuous mass. The acceleration of the ball produced by the hand is given by Newton's Second Law of Motion: *the acceleration of a mass equals the force acting on it divided by the mass.*

[A scientifically more useful statement of the law is based on the concept of 'momentum'. For a discrete mass this is the product of the mass and its velocity. Velocity is a vector quantity because it possesses both speed and *direction*. *The rate of change of momentum of the mass equals the applied force.* This concept of momentum and the fact that it is changed by a force is very important and features frequently in the following sections on aerodynamic forces.] In order to analyse the subsequent motion of the centre of mass of the ball, we do not need to model all the internal molecular interactions involved in its distortion – they cancel out. We just need to know the volume of the ball and the average density of its material, which give the mass, and the force applied to the ball, which includes gravitational force and aerodynamic drag.

The conceptual difficulty in extending this ball model to a fluid element is that fluid molecules not only interact but *mix* with those of other neighbouring elements. At any instant of time, the solid ball model and Newton's Second Law are sufficient to allow the motion of the ball to be analysed. But in order to analyse the movement of a fluid element over time it is necessary also to model the passage of molecules into and out of the imaginary envelope that we use to define the boundary of our element. For example, if the rate of inflow of molecules is greater than the rate of outflow, the average density of the element increases with time, as does the mass.

The principle of 'Conservation of Mass', which expresses the fact that mass lost from one element must be gained by neighbours, must be applied together with Newton's Laws of Motion. We have seen that air pressure is a manifestation of the average kinetic energy density of air molecules. In *microscopic* terms, molecules interact across the boundary of our element. If the kinetic energy density of the molecules in an element neighbouring a contiguous element are more energetic on one 'side' of the latter element than on the other, they cause a net flow of the molecules of the latter away from the more energetic neighbour. According to the alternative *macroscopic* model, if an element of air is subjected to different pressures on opposite sides, it is subject to a net force and therefore accelerates in the direction of that force. Since pressure in free air varies smoothly over distance it is more precise to relate the pressure difference across an element to the spatial rate of change of pressure, or *pressure gradient*, through the element.

The significance of pressure gradients in the air (and in all fluids) is illustrated by the behaviour of a helium-filled balloon inside a moving car. The balloon sticks to the roof of the car because it is subject to the same atmospheric pressure gradient caused by gravity to which all the surrounding air is subject, as explained in Chapter 1. But it is lighter than the air that it displaces, and therefore its weight is less than the upward force imposed upon it by the surrounding air. When the car accelerates, the occupants and contents would move towards the back of the car if not constrained by contact forces between them and the seats. However, the balloon moves towards the *front* of the car. For the air in the cabin to move with the car (that is, share the car's acceleration), there must be a gradient of air pressure from higher at the back to lower at the front. This pressure gradient also acts on the balloon, but the helium balloon has less mass than an equal volume of air, and so it suffers greater acceleration than the air and moves towards the front of the cabin. We may generally ignore the effect of gravity, and the vertical pressure gradient that it imposes on the air, on small-scale airflows that occur near the surface of the earth, although it has profound effects on the large scale airflows that are the concerns of meteorologists (see Chapter 5).

Acceleration is commonly thought of as a change of *speed*. This definition is, however, incomplete. Acceleration is defined as a change of *velocity*. Velocity is a *vector* quantity, which means that it has not only magnitude (speed), but also direction. Vectors are graphically represented by arrows, the length of which is proportional to speed. *Change in either speed or direction, or both, constitutes acceleration.*

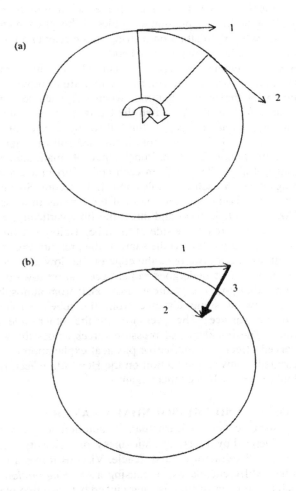

Fig. 3.1 Acceleration in a rotating system

Consider the velocity vector (1) of a point on the tread of a rolling tyre (Fig. 3.1(a)). Represent it by an arrow of length proportional to the speed and appropriate tangential direction. Now, consider the velocity vector of this point (2) a little later in time (the angle of rotation has been made fairly large for graphical clarity). The *length* of the arrow (speed) is the same but its *direction* is different. Bring the origin of the second arrow back to coincide with that of the first as in Fig. 3.1(b). The change in velocity vector, which is proportional to the acceleration vector, is given by the arrow (3) joining the tip of the first arrow to that of the second. If the angle of rotation is assumed to become very small (approaches zero), the direction of the acceleration passes through *the axle*. This construction can be made for any curved trajectory. Thus, a mass moving along a *curved path* at *constant speed* is accelerating towards the *centre of curvature* of the path and requires the action of a force in that direction to do so. Common examples of this phenomenon are the force exerted by a person whirling a stone on a string in a circle; and the force acting on the passengers by a vehicle that navigates a bend.

When a fluid element moves on a curved path there must be a fall of pressure across the flow towards the centre of curvature to provide the accelerating force, explaining why the helium balloon, when subjected to the same pressure gradient as the air, moves to the *left* when the car takes a left hand bend. You may demonstrate this important feature of fluid flow by means of a simple home experiment. Take the cardboard tube from a finished roll of cling film or cooking foil. Cut off a length of about 250 mm. Punch a pair of small, diametrically opposed holes at a distance of about 25 mm from each end. Thread a 2 m length of cotton thread (not string) through each pair holes and tie to secure. Suspend the tube in a horizontal position by fixing the other ends of the threads to a high bar such as a curtain pole. Now take a hair dryer and direct the flow vertically, either upwards or downwards and *slowly* approach the side of the tube. Before reaching the side of the tube, the pressure in the dryer flow is the same as the pressure everywhere else in the room – namely, atmospheric. But once the edge of the flow is forced to bend round the curvature of the tube, it undergoes radial acceleration towards the axis of the tube. Therefore a radial pressure gradient must exist; from atmospheric pressure in the non-diverted flow at some distance from the tube's surface, to less than atmospheric on the surface. The pressure on the other side remains at the atmospheric value. The imbalance of pressure forces across the width of the tube causes the observed effect. An alternative physical explanation of the behaviour of the tube is presented below in the section on the Bernoulli effect. [Keep thread and the suspended tube – you will need them again.]

3.3   VISCOSITY AND THE BOUNDARY LAYER

It may surprise you to learn that, in addition to pressure gradients, air motion is also influenced by a form of 'stickiness' or 'viscosity' such as we usually associate with liquids like motor oil and treacle. Viscous forces arise when adjacent layers of fluid have different speeds, so causing a *shearing motion* as illustrated by Fig. 3.2 in which a thin layer of fluid is sandwiched between two plane surfaces, one of which is sliding past the other: the fluid 'sticks' to each of the surfaces. The fluid is sheared, which creates shear forces that act so as to reduce the speed difference. Newton postulated that the viscous force per unit area (stress) is proportional to the rate of shear, which is given by the velocity difference (V_1 - V_2) divided by the gap width d. The constant of proportionality is the 'coefficient of dynamic viscosity'.

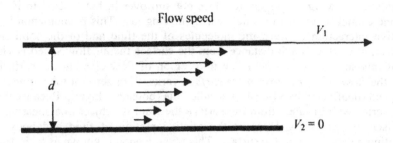

Fig. 3.2 Shear flow

In the case of liquids, viscosity is due principally to attractive forces between molecules; these are not, however, strong enough to enforce the maintenance of shape, as they do in solids. Heating a liquid energises the molecules and makes it more difficult for them to stick together, thereby decreasing the viscosity. But gas molecules are, on average, so far apart, and interact so infrequently, that one would think that air would be effectively non-viscous (inviscid). However, this is not so. The reason can be illustrated by considering the analogy presented by two small boats travelling in the same direction on parallel courses, with one going faster than the other, which is a crude an analogy of two adjacent elements within the shearing motion of a fluid. If a person in the faster boat throws a sack of sand into the slower boat while passing it, the latter will speed up. If the bag is thrown from the slower boat into the faster boat, the latter will slow down. These effects are a consequence of the 'Principle of Conservation of Momentum'. Now consider two adjacent layers of smoothly flowing air, one flowing faster than the other. As we know, air molecules fly around in all directions; some will leave the faster layer to jump into the slower layer, and vice versa, in a process known as diffusion. The effect is to tend to reduce the speed difference between then layers. Through molecular 'collision', it also converts a proportion of the kinetic energy of directed molecular motion (organised flow) into random molecular motion (heat). It is thus a source of loss of flow energy. Increasing air temperature increases molecular speed, and therefore *increases* air viscosity. We shall see later in this chapter that the process of mixing on a larger scale, called 'turbulence', transfers kinetic energy between fluid elements in a far more efficient manner.

Remarkably, the relatively weak actions of viscosity are crucial to the behaviour of air flowing over moving bodies. *Without viscosity, aircraft could not generate lift and fly.* It controls the drag that their propulsion systems have to overcome. It also contributes a substantial portion of the drag experienced by land vehicles. In the nineteenth century, scientists predicted theoretically that the drag on moving bodies would become negligible at very high speeds; on the other hand, practical hydraulic engineers knew that it actually increases. This was known as the d'Alembert paradox. Jean le Rond d'Alembert was a French mathematician, the son of a nun.

This paradox was solved in the early twentieth century by Ludwig Prandtl, a German physicist. He observed that in the flow of fluid over a solid surface, fluid *immediately adjacent* to the surface does not slip over it, but 'sticks' to it. This is why one cannot blow dust off one's car by driving fast. This phenomenon involves attractive interaction between the molecules of the fluid and of the solid surface. However, at a rather small distance from the surface the air flow speed is virtually free of this surface 'stickiness'. The thin region of flow close to the surface over which the flow speed *relative to the surface* varies from zero (at the surface) to the free speed (unaffected by viscosity) is called the 'boundary layer'. Because the flow speed varies with distance from the surface the fluid is subject to a shearing action and viscosity acts so as to resist the relative motion of fluid and body, hence generating a drag force on the surface. This phenomenon is known as 'skin friction'. The drag force on a body moving through the air does negative work on the body and positive work on the air. The energy given to the air is irrecoverably dissipated into random molecular motion (heat).

Skin friction generates the fiery streaks of meteors (shooting stars) that mostly burn up before reaching the Earth's surface. The atmosphere protects the Earth's surface from suffering the fate of the pock-marked moon, and allows life to exist. Space vehicles returning to Earth are protected by an ablative layer that is heated to enormous temperatures by skin friction, melts, and blows away to dispose of the otherwise deadly heat. As we shall see a little later, this is only one of the ways in which viscosity affects the forces on a body moving through a fluid. In the free flow outside the boundary layer, viscosity exerts little influence.

3.4   LAMINAR AND TURBULENT BOUNDARY LAYERS

The boundary layer on the surface of a long, thin body moving *slowly* through still air (or, equivalently, air moving *slowly* over a static body) starts at the 'nose' with negligible thickness and thickens slowly with distance along the body by means of molecular diffusion, which was explained above as the source of gas viscosity. The flow is steady and smooth and the trajectories of fluid particles may be defined by 'streamlines' which are tangential to the local flow directions. The variation of flow velocity with *perpendicular* distance from a surface is known as the 'velocity profile'. It is parabolic and corresponds to the line joining the tips of the arrows in Fig. 3.3(a). This steady form of boundary layer is termed 'laminar'. Small random disturbances of this smooth flow pattern, such as those generated by roughness of a body's surface or turbulence in incoming flow, are rapidly damped out by viscosity which acts so as to suppress random fluid motion: the boundary layer is stable. As the product of airspeed and distance from the 'nose' increases, the control of such disturbances by viscosity becomes weaker. Eventually the whole flow becomes unstable and superimposed on the streamwise flow there exist random fluctuations of flow velocity in the form of regions of circulating flow, known as 'eddies', that grow, mix and die as new ones arise. [A similar process is seen in the upward flow of smoke from a cigarette.] This is known as boundary layer 'transition' into a 'turbulent' boundary layer.

Fluid dynamicists characterise flows over solid bodies and within conduits in terms of a non-dimensional parameter called 'Reynolds Number', named after Osborne Reynolds who studied the transition from laminar to turbulent flow in pipes in the 1880s. Reynolds number is a measure of the ratio of so-called 'inertia forces'

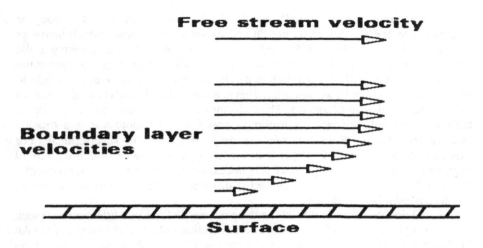

Fig. 3.3(a) Laminar boundary layer velocity profile

to viscous forces. Inertia forces are given by the product of the fluid particle accelerations and the fluid density; they may be thought of as a measure of the difficulty of stopping the velocity fluctuations that the viscous forces oppose. The Reynolds Number (*Re*) of a particular flow over a particular body is given by $Re = Vd\rho/\mu$, where V is the speed of the oncoming flow (relative to the body), d is a length that characterises the size of the body, ρ is the fluid density and μ is the coefficient of dynamic viscosity. At values of *Re* much less than a 'critical' value that is associated with any particular geometric form of body, the flow is controlled by viscosity. At values of *Re* much greater than the critical value, viscous forces exert little influence; the flow is dominated by unsteady turbulent velocity fluctuations and associated distributions of surface pressure. The importance of Reynolds number is that it determines whether any particular form of flow is controlled by fluid viscosity or not. This knowledge is vital to the design of low-drag vehicles. The following are examples of approximate Reynolds number in air at sea level. Falling raindrop: order of 100 (based upon diameter): Bumble bee in straight flight : 7500 (based upon length): Tennis ball served at an initial speed of 100 kph: 106,000 (based upon diameter): Golf ball struck at an initial speed of 250 kph: 180,000 (based upon diameter): Typical saloon car at 90 kph; 6,750,000 (based upon length): Airbus 380 at cruise: 1,340,000,000 (based upon fuselage length)

The critical Reynolds Number for flow over a sphere is about 100,000. At very low Reynolds numbers of the order of 1000 or less, the boundary layer remains laminar and the drag is dominated by skin friction. At sub-critical Reynolds numbers of the order of 10,000, the boundary layer remains laminar until it separates (leaves the surface of the body) in the plane of maximum width to form a wide wake and high drag. At much higher, super-critical, Reynolds numbers the boundary layer becomes turbulent before separating and doesn't separate until further round the downstream arc of the surface to form a thinner wake, and the drag decreases (see also below).

The maximum (terminal) speed of fall through the air of a body is determined by the equality of its weight and its aerodynamic drag, which increases with speed. For a sky diver, this speed is about 125 mph (200 kph). Looking at the table of typical Reynolds numbers above, we can deduce that the maximum (terminal) speed of fall of a raindrop is determined by the equality of its weight to the drag imposed on it by air viscosity. This is an important factor in cloud dynamics as discussed further in Chapter 5. The trajectories of tennis balls, footballs, cricket balls and baseballs are rather sensitive to speed, surface roughness and rotation because their typical Reynolds numbers range from sub- to super-critical. For example, the trajectory of a cricket ball bowled by a fast bowler (initially 85 to 90 mph) can sometimes be straight until it bounces, slows and rotates, and subsequently be strongly curved. This general subject is discussed in detail below in the section entitled 'Bend it like Beckham'.

Fluid flows over bodies of the same geometric form but *different sizes* (such as a post supporting a road sign and a circular-section industrial chimney stack) take the same *form* if the Reynolds numbers of the two flows are the same. This poses an interesting problem for wind tunnel testing of small models of aircraft. The Reynolds numbers of the model and the full scale aircraft should, in principle, be equal for the flows to take the same form; that is to say, for the flow over the model to simulate that over the aircraft. However, if air is used in the wind tunnel, ρ and μ are much the same in both cases. Therefore, the speed of flow over the model should be greater than that over the aircraft by a factor equal to the inverse of the model scale, which is typically of the order of 1/50. Suppose the aircraft flies at subsonic speeds up to 800 kph; the model should be tested at speeds up to 40,000 kph! It is obviously impossible to generate such speeds; anyway, this speed is hypersonic and shock waves would dominate the model flow. The problem is overcome by the cunning application of rough surface strips of boundary layer 'trips' near the leading edges of wings and tail which destabilises the laminar boundary layer and turns it turbulent where it would naturally do so on the full scale aircraft. The flows over the model *outside the boundary layers* at the full scale flight speeds then closely simulate those over the aircraft at the same speeds. This is very lucky for aeronautical engineers. In cases of supersonic flight, the second non-dimensional parameter that must be the same for model and full scale aircraft is the 'Mach Number' (M). This is the ratio of the flight speed to the speed of sound (see Chapter 4). The speed of sound in air is proportional to the square root of absolute temperature, so this requirement does not pose a difficult practical problem of flight simulation by wind tunnel models.

The boundary layer on a *smooth* flat plate remains laminar over a streamwise distance corresponding to a Reynolds number of about 100,000, and is definitely turbulent beyond a streamwise distance corresponding to a Reynolds number exceeding 2,000,000. As indicated above, at Reynolds numbers less than about 100,000, the boundary layer on a *smooth* sphere remains laminar until it separates from the equator of the sphere to form a wake that largely controls the drag. But, at higher speeds the boundary layer becomes turbulent before separation. Since a turbulent boundary layer resists separation far more strongly than a laminar boundary layer, the separation line moves further to the rear of the sphere, the wake gets thinner and the drag decreases substantially. We shall see the importance of this to golfers later in the chapter.

Fig. 3.3 (b) Turbulent boundary layer flow

Fig. 3.3(b) shows an instantaneous snapshot of a turbulent boundary that is flowing from left to right and is separating from the surface at the extreme right hand side. The turbulent boundary layer thickens more rapidly than the laminar boundary layer because it involves mixing of fluid elements having different speeds on a very large scale compared with the molecular scale involved in diffusion, in a rough analogy to the difference between forced and free convective heat flow described in the section on heat transport in Chapter 1. It can thus 'pull in' or 'grab' the energy of high-speed elements from the outer edge of the boundary layer and pass it down to the elements close to the surface to assist them to overcome the opposing viscous forces. The unsteady pressures and flow velocities associated with turbulent boundary layers can have both harmful effects, such as structural damage and noise, and beneficial effects, as in the enhancement of heat exchange in industrial processes and the dispersal of atmospheric pollution.

3.5   BERNOULLI'S EQUATION
We now digress briefly to introduce a very important relationship between fluid pressure and flow speed. Imagine a tapering tube through which a fan at the smaller diameter end draws air. As the air passes through the decreasing cross-sectional area of the tube the air speed increases so that the principle of conservation of mass is satisfied. [We exploit this effect when pinching the open end of a garden hose to speed up the outflow and therefore spray further; but the rate of flow of water mass (or volume) is the same at all points of the hose, irrespective of kinks or bends or pinching.] As the air passes down the tapering tube, an increasing proportion of the kinetic energy of random molecular motion, which produces pressure, is converted into kinetic energy of ordered motion (along the tube). Thus,

air pressure falls as the flow speed increases. (Strictly speaking, the pressure also falls to a small extent because of the frictional action of viscosity in the boundary layer on the wall of the tube: but we shall neglect this for the moment.)

This phenomenon, in which the *pressure falls as the speed increases* is an example of the principle of conservation of energy and is expressed by an equation attributed to the Swiss scientist Daniel Bernoulli. One example of this effect is seen in the upward bulging of the hood of a soft-top car as it displaces air that speeds up as it flows over the concave surface. Large umbrellas can be difficult to control in strong winds partly because the flow is 'caught' and slowed down by the concave under-surface, which increases the pressure; and partly because it speeds up over the curved top surface, which reduces the pressure. The resultant difference of pressure generates force. You may observe that many large umbrellas have flaps placed over a central hole; they exclude the rain but allow air to flow from below to above the canopy, thus 'short circuiting' the pressure difference. This phenomenon also provides an alternative explanation of the reduced surface pressure on the side of the cardboard tube exposed to the flow from the hair dryer used in the home experiment described above; in navigating the curve of the dryer flow, the flow is effectively 'squeezed', and speeds up.

A more vital example of the Bernoulli effect is the generation of sound by the human vocal cords which are rather like a pair of thick elastic bands lying together, stretched across the trachea (wind pipe). By increasing the pressure in the lungs, we can open them. Air speeds up through the gap, which lowers the air pressure between them and pulls them together. Then the air pressure builds up and opens them again; and so on. Thus they vibrate, releasing a sequence of puffs of air which create sound. Open your mouth and say 'aaaahhh'. This is the motor that drives your voice. Tighten up or loosen the vocal cords to generate a higher or lower pitch. Or you can maintain a more or less constant pitch and change the shape of your oral cavity to create speech sounds such as the vowels: but more of this in Chapter 4.

All forms of body in an airflow have points towards their front at which the flow comes to a halt *relative to the body*. This is called a 'stagnation point'; the pressure reaches a maximum here, as indicated by Bernoulli's equation. This is exploited by a flow speed measuring device, seen in Fig. 3.4, which is commonly used by aircraft and boats; it is called a 'Pitot' tube, after the Italian/French aeronautical engineer who invented it. You can see it protruding from the wingtips of aircraft. The air is brought to a halt at the entrance of the central tube, which increases the local pressure and flows freely over the outer tube with no change of pressure. The difference of pressures is proportional to the square of the flow speed (or the vehicle speed relative to the fluid).

3.6   HOW DO VISCOSITY AND THE BOUNDARY LAYER AFFECT AIRFLOW?

If air lacked viscosity (inviscid), it would slip freely over bodies of all shapes and the flow would close cleanly behind the body as shown by the streamlines in the theoretical case of a circular cylinder in Fig. 3.5. The forces on these bodies would then arise solely from the action of the air pressure on the surface

Fig. 3.4 Pitot tube

Although the air at first speeds up and then slows down as it passes over the bodies, and therefore the pressure falls and then rises again, neither these bodies, nor any others of any shape, would experience drag. However, all fluids exhibit viscosity, and boundary layers form on the surfaces of all bodies. They remain thin over the forward part of a body as the flow speed increases and pressure falls. However, towards the rear of a body, the airspeed starts to fall as the flow 'attempts' to close up, and the pressure begins to rise. Consequently, the rising pressure tends to oppose and slow the flow; this is called an 'adverse pressure gradient'. The greatest retardative effect is 'felt' by the low speed, low energy flow close to the surface within the boundary layer, which thickens and separates from the surface, as shown diagrammatically in Fig. 3.6 and exemplified by the flow over a cylinder, shown in Fig. 3.7. *Boundary layer separation is an extremely important phenomenon that has widespread effects on the flow over bodies of every type.*

As previously explained, turbulent boundary layers are much more energetic than laminar boundary layers because the large eddies 'grab' energy from the external flow and pass it down to the sluggish flow near the surface, which is a much more efficient process than that of molecular diffusion that drives the laminar boundary layer. Consequently, turbulent boundary layers separate less readily than laminar boundary layers; however, they will separate if the pressure gradient is sufficiently adverse, as seen at the extreme right hand side of Fig.3.3(b). This has important consequences for wake generation, drag and ball games, as illustrated later.

Fig. 3.5 Ideal inviscid flow over a cylinder

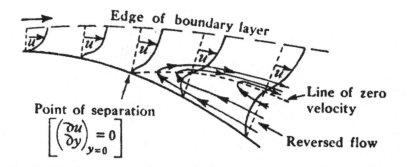

Fig. 3.6 Boundary layer separation and flow reversal

Fig. 3.7 Boundary layer separation from a cylinder: flow left to right

The separation of the boundary layer causes the formation of a wake (low flow energy) behind a body. The flow pattern does not speed up and close as it would in the absence of viscosity, and the pressure does not rise to the value it would have done in the absence of viscosity. The resulting lack of balance of pressure between front (where the pressure reaches a maximum at the stagnation point) and the rear produces a drag force on the body. There is also a component of drag due directly to the action of the boundary layer skin friction on the body's surface. The crucial point is that the action of viscosity in generating a boundary layer *indirectly* creates drag by causing an imbalance of surface pressure.

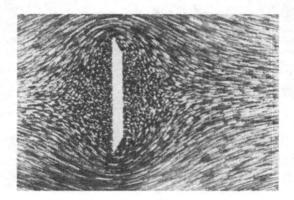

Fig. 3.8 (a) Low speed flow around a plate

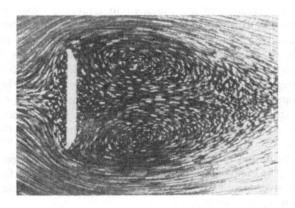

Fig. 3.8 (b) Medium speed flow around a plate

Fig. 3.8 (c) High speed flow around a plate

Fig. 3.9 Flow over a modern car

Streamlined bodies such as wing sections that are thin and come to a sharp edge at the downstream end are designed to minimise flow separation and the consequent pressure contribution to drag. The drag of all non-streamlined (bluff) bodies is dominated by the 'pressure drag'. A dramatic effect of separation of flow from the sharp edges of a thin plate placed perpendicular to the flow is shown in Figs. 3.8 (a), (b) and (c). The airspeeds are in the ratios 1:4:1000. Note how strong circulating flows (vortices) are formed behind the plate at the two higher speeds. The vortices separate from the plate and flow downstream only at the highest speed to form a vortex street. This is why barriers employed to protect plants from damaging winds are not solid, but consist of a mesh that generates many small, harmless vortices whose energies are rapidly dissipated by viscosity. The generation of pressure drag by boundary layer separation is why the rear-end geometry of a car is more influential in controlling the drag force than that of the front end. Fig. 3.9 shows the flow over a modern car in a wind tunnel, with separation at the rear of the roof, revealed by smoke streams.

It might be thought that, compared with a smooth ball, the dimples on a golf ball would adversely affect the flight distance. However, this is not the case. Golf balls slow down by a factor of about three during flight. Without dimples, the boundary layer would be laminar for much of the flight duration, and would separate near the mid-plane of the ball, creating a wake of similar diameter to the ball and a large pressure drag. As mentioned above, turbulent boundary layers are more capable of overcoming adverse pressure gradients. The dimples trip (destabilise) the laminar boundary layer growing over the forward surface and cause transition to a turbulent boundary layer which separates further around the periphery of the ball and creates a thinner wake and less drag. It is said that the 'fluff' on new tennis balls plays the same role. A similar phenomenon is employed on some aircraft wings to reduce pressure drag. Arrays of small, upstanding tabs (vortex generators) set at an angle to the airflow are distributed over the forward upper surface of the wings. These stimulate mixing of the boundary layer flow with the high speed free flow and thereby energise the former and suppress boundary layer separation.

The surprising indirect influence of viscosity and the associated boundary layer is the flow phenomenon that Prandtl revealed as a principal cause of drag except in flows of very low Reynolds number (very small objects such as plant spores or very low speeds or very viscous fluids such as treacle). In the cases of bluff bodies the separated boundary layers are unstable and roll up into 'vortices' as seen in Figs. 3.8 and 3.19. These can stimulate damaging oscillations of cables, chimney stacks and can also create singing sounds. We shall meet vortices again later in this chapter.

Viscosity and the boundary layers that it causes is not only responsible for a substantial part of the drag force on bodies, but is also crucial to the 'lift' generated by aircraft wings of which the streamwise cross-sections are termed 'aerofoils' (USA: 'airfoils'). Lift is the component of flow-generated force on a body that is directed perpendicular to the direction of its motion relative to the oncoming fluid. In order to produce a vertical lifting force, an aerofoil moving horizontally through the air must induce *downward velocity* and, therefore, *downward momentum*, in the air. The rear-mounted 'wings' on racing cars induce upward velocity to generate increased load, and therefore increased friction, on the rear wheels. This is in accordance with the principle of 'Conservation of Momentum', which is another way of expressing Newton's second law of motion. You can experience this effect (if not driving) by holding your flat hand out of the car window in a horizontal plane and then *gradually* tilting its leading edge upwards – *be careful not to do this in traffic*! The effect of changing the direction of a stream of fluid can also be readily observed by laying the end of a garden hose in a loop and then turning on the water. As explained above, the passage of the water round the loop accelerates it towards the centre of the loop and the reaction force straightens the hose.

3.7   THE DEVELOPMENT OF AEROFOILS FOR WINGS
It would appear that any plate-like object set at an angle to a wind (angle of incidence, or attack) will experience a lift force irrespective of the presence or absence of a thin boundary layer. However, theoretical analysis of ideal *inviscid* flow around a very thin flat plate shows that the direction and speed of the flow (that is, the flow velocity vector) downstream of the plate are the *same* as that of the oncoming flow. Momentum is a vector quantity equal to the product of the mass and the velocity vector of the mass. As explained in a previous section, momentum is changed by an applied force. Since the vertical component of the momentum of the airflow is not changed by the presence of the plate, there is no net vertical force acting on the plate or on the flow. As shown in Fig. 3.10(a), the flow takes a symmetric form, with high pressure stagnation points S' and S on upper and lower surfaces. Although the inviscid flow round an aerofoil, seen in Fig. 3.10(b) does not possess the same symmetry as that around a plate, there is similarly no net change in direction of the flow streamlines. The fluid momentum undergoes no net change between upstream and down stream, and both lift and drag are zero.

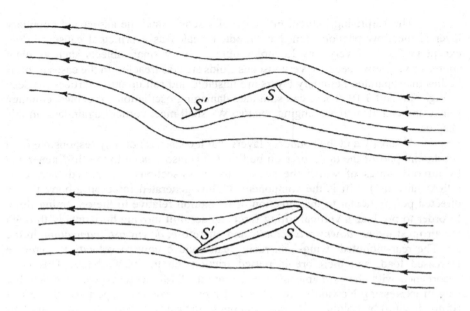

Fig. 3.10(a) & (b) Inviscid flows over a flat plate and an aerofoil at incidence

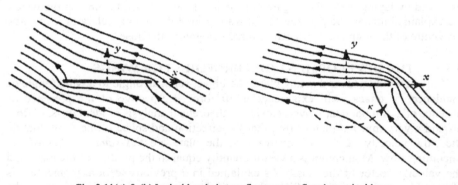

Fig. 3.11(a) & (b) Inviscid and viscous flows over a flat plate at incidence

So why is the lift not actually zero in practice? This is where the viscosity has a crucial influence. When the plate starts to move forward through the air, the flow *initially* takes the form shown in Figs. 3.10(a) and 3.11(a) given by inviscid theory, because the boundary layers haven't had time to establish. Stagnation points where the pressure is highest exist on both lower and *upper* surfaces. Air approaching the trailing edge along the *lower* surface has to make an upward U-turn around the sharp trailing edge. The extreme curvature of the flow streamlines indicates that radial acceleration of the flow is very large and directed towards the edge. We know from the earlier discussion of curved flow that the associated pressure at the trailing edge must therefore be very low indeed. In contrast, the pressure at the rear stagnation point on the *upper surface* is high. Therefore, there is an extremely large adverse pressure gradient from the trailing edge to the rear stagnation point which opposes, retards and reverses the local surface flow which is 'attempting' to get round the trailing edge. The net result is that a region of the, by now, established boundary layer flow separates to form a rapidly growing 'bubble' of low energy flow on the upper surface. The stagnation point to the upper surface is pushed back by the large pressure gradient from the rear stagnation point to the trailing edge as can be seen by comparison Fig. 3.11 (b) with the inviscid flow shown in Fig. 3.11 (a). The flow leaves the trailing edge tangentially, unlike the inviscid case, and thin 'wake' of low energy fluid is formed downstream of the plate through the merging of the upper and lower surface boundary layers. The stagnation point on the forward part of the lower surface remains largely intact and the resulting imbalance of pressures between upper and lower surfaces produces lift and drag. Note that the directions of the flow streamlines approaching and leaving the plate in Fig. 3.11(b) are different: the vertical component of fluid momentum flux has been changed by the presence of the plate and the effect of viscosity, giving rise to lift.

The generation of lift can also be explained in terms of the curvature of the flow streamlines which necessarily implies the presence of a transverse pressure gradient. As seen in Fig. 3.10(a) and 3.11(a), the streamlines passing upwards around *both the leading and trailing edges* are strongly curved; therefore the local surface pressures are below the pressure in the undisturbed approach flow. A fairly similar flow pattern around the *leading* edge appears in Fig. 3.11(b), producing low pressure on the forward part of the upper surface, but the flow leaves the trailing edge tangentially and does not balance the upward force of the former. There is a lack of symmetry of curvature of the flows underneath and above the plate in Fig. 3.11(b) and, the pressure under the plate exceeds the undisturbed pressure and is higher than on the upper surface. The resulting force imbalance produces lift.

The flat plate does not make a good aircraft wing because its ratio of lift to drag forces is low, and the flow over the upper surface breaks away suddenly from the leading edge at moderate angles of incidence (angle to the direction of oncoming flow) causing a catastrophic loss of lift called 'stall'. This failing was realised by aeronautical pioneers of the 19th Century who experimented with gliders of various sizes. These pioneers mostly used thin curved (cambered) sections which can truly be called 'aerofoils'. They constructed whirling arms and wind tunnels to discover the optimum curvature. As we have seen earlier, curvature of a flow path requires a pressure gradient perpendicular to the flow. So, relative to a flat plate, the curvature increases the flow speed and reduces the pressure over the upper surface, and increases the pressure on the lower surface, thereby increasing the lift. Although thin cambered sections produce a high ratio of lift to drag at small angles of incidence to the oncoming flow, the thin sections are structurally very flexible and require extensive drag-producing bracing for shape maintenance under load. The thin leading edge also causes severe stall behaviour. The pioneers should have looked more closely at the cross-sectional shape of the wings of some of the more aerodynamically efficient gliding birds, such as the albatross, that have a fairly thick, rounded leading edges as well as camber.

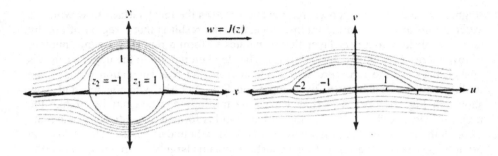

Fig. 3.12 Mathematical transformation of a flow around
a cylinder into flow around an aerofoil

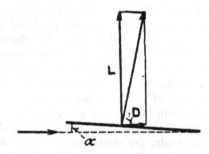

Fig. 3.13(a) Lift and drag forces on a flat plate at angle of incidence α

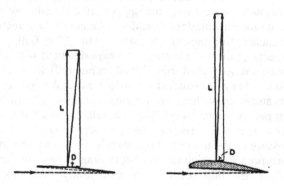

Fig. 3.13(b) and (c) Lift and drag forces on a thin cambered plate and an aerofoil
at angle of incidence α

The answer to both the structural and stall problems was the introduction of specially shaped thicker aerofoil sections theoretically developed on the basis of a technique developed by Nikolai Joukowsky, a Russian mathematician. As illustrated by Fig. 3.12, this transforms the well-known flow over a cylinder into an aerofoil shape on which the pressure distribution is easily calculated. Typical lift and drag force vectors of flat plates, cambered plates and simple aerofoils are plotted in

Figs.3.13(a), (b) and (c), respectively. Note the increasing ratio of the lift force to drag force. Modern wings are designed to maximise the ratio of lift to drag and are well behaved over a large range of speed and angle of incidence. However, there is a limit to the maximum angle of incidence for which boundary layer remains attached to the upper surface of an aerofoil section. Increase of the angle beyond this limit causes separation of the flow, as seen in Fig. 3.14, loss of lift and the very dangerous state of 'stall'. Fixed aerofoil sections alone cannot deal with the demands of modern aircraft performance auxiliary, deployable extensions to the leading and trailing edges are installed to allow safe flight at angles of incidence much higher than a fixed section would allow. You can witness this deployment as your plane takes off and as it approaches the landing strip. These allow aircraft to generate high lift at relatively low speed for short take-off distances and to descend rapidly and safely for short landing distances. Fig. 3.15 shows a modern airliner landing with it all 'the washing hanging out'. Some birds deploy similar auxiliary wing extensions as described later in this chapter

Fig. 3.14 Flow over a stalled aerofoil

Fig. 3.15 Airliner coming into land with flaps and undercarriage down

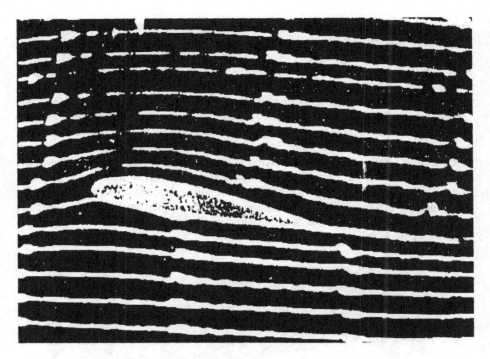

Fig. 3.16 Smoke streamlines over an aerofoil at incidence

On the basis of the common (but false) assumption that neighbouring pairs of air particles that separate at the leading edge, one flowing above the aerofoil and one flowing beneath, meet again at the trailing edge, the flow speed over the top must be the greater and hence the pressure is lower on top than below (Bernoulli), thereby generating lift. The relative positions of the blobs (puffs) injected into the smoke-loaded streamlines shown in Fig. 3.16, that were generated simultaneously in the uniform flow some distance upstream (to the left) of an aerofoil in a wind tunnel, give the lie to the assumption of equal transit times. The fact that aircraft can fly upside down demonstrates that the different surface curvatures are not essential to the generation of lift; even an aerofoil of symmetric section generates lift. The vital element is that vertical (downward) component of the momentum flux of the air must increased by the presence of the aerofoil. The differential pressures on the upper and lower surfaces can also be understood in terms of the differential curvature and associated radial accelerations of the upper and surface flows, as explained above.

3.8   SAILS

The sails mounted on yachts form aerofoils and produce horizontal force by means of the same basic mechanism that aircraft wings produce lift. The 'spinnaker' is the exception because it is usually deployed to produce drag to sail before the wind. However, the characteristics and performance of most sails are quite different from those of the aircraft wings because, except for fixed 'wing' vessels, the shape and relative dispositions of the sails in relation to the wind direction and strength are manipulated by the crew to optimise the vessel's performance in relation to wind and wave conditions. In particular, the curvature of a sail is controlled by the distribution of the aerodynamic force acting on it, its material properties and design, its edge fixtures and the lengths and tensions of the sheets (lines) as adjusted by the crew to set the relative dispositions of the two or more sails. The progress of a yacht is the result of the sums of the actions of two major forces: the sail forces on the mast (s) and the hydrodynamic forces on the hull. By means of making appropriate adjustments to the configuration(s) of the sails, and making appropriate manoeuvres (tacking), yachts can make progress against the wind.

3.9   VORTICITY AND VORTICES

Fluid flow is commonly depicted in terms of lines called 'streamlines' that indicate the paths of particles; these lines maybe straight or curved. However, fluid elements may also rotate. This feature plays a vital role in various types of flow, many of which are of practical importance. We have seen that spatial variations of pressure (pressure gradients) cause accelerations of fluid elements that may involve changes of speed or changes of flow direction, or both; but pressures cannot cause element rotation. It requires the action of viscous 'shear' forces (that exist in flows such as laminar the boundary layer seen in Fig. 3.3(a)) acting on fluid elements to cause them to rotate as they move. The measure of this rotation is defined as 'vorticity' which is numerically equal to twice the rate of rotation. Vorticity is a fundamental feature of fluid flow that is usually produced by interaction between viscous fluid flow and solid surfaces, as in boundary layers, or by the interaction between two airflows of different speeds, such as flow emerging from a pipe in the atmosphere, which creates a shear layer. The generation of vorticity is essential to the generation lift and drag by a wing. Vorticity obeys certain 'laws', such as that of persistence (or long life) which are fundamental to aerodynamic theory and to the understanding of many forms of fluid flow; but these aspects generally lie outside the scope of this book and more technically minded readers may wish to consult textbooks on fluid dynamics for further elucidation.

A vortex (not to be confused with 'vorticity') is a form of flow that involves flow along a circular or spiral path. One of the most the most commonly observed examples is that generated in a cup of liquid by stirring it with circular movements of a spoon. If, after a short time, the spoon is withdrawn, the fluid continues to rotate like a rigid cylinder, except for the thin boundary layer at the walls of the cup. The flow speed is *proportional* to the distance from the centre and the fluid level is lowest at the centre, as seen in Fig. 3.17 in which depth of the well is greatly exaggerated. As we have seen, motion of fluid particles along a curved path requires a transverse pressure gradient. The depth of fluid above any submerged horizontal plane in the rotating fluid increases towards the walls as does the hydrostatic pressure, so the pressure gradient provides the necessary radial force to allow the fluid to follow circular paths; this type of vortex is called a 'forced' vortex. In following a circular trajectory, the individual fluid elements rotate about their centres once per revolution and therefore this type of vortex is termed 'rotational'.

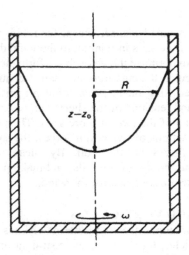

Fig. 3.17 A rotating fluid exhibits a parabolic surface shape

Another commonly observed vortex is that produced by the outflow of water from a bath. In this case, careful observation of floating particles shows that the flow at a distance of about twice the radius of the plug hole constitutes a 'free' vortex in which the individual fluid elements particles follow a circular trajectory but *do not rotate* about their individual centres, rather like the gondolas on the London Eye. [Try floating a few short pieces of match stick, or a plastic duck, at some distance from the plug.] This form of flow is, strangely enough, termed 'irrotational' because it has no vorticity. The circular flow speed is *inversely* proportional to the distance from the centre of the vortex, and therefore increases rapidly towards the centre. The water level falls towards the vortex centre, as does the pressure which sustains its centripetal acceleration. At some point the pressure falls below the vapour pressure of the water and a central funnel of vapour-saturated air forms. On approach to the edge of the plughole, the rotational motion is combined with the outflow, viscosity starts to have an influence and particles start to rotate individually and the water level plunges down into the plug hole. As explained in Chapter 5, the direction of rotation is determined by the initial state of the water, not by the location on the Earth's surface.

Vortices may be seen as dimples on the surface of slowly flowing rivers. They may also be conveniently observed in the shadows that they form on the bottom of a bathtub or other light coloured container in which the water is illuminated by bright overhead light. Hold a ruler vertically so as to penetrate the water surface and hold it an angle of incidence of about $30°$ to the direction of motion while you move it forward to generate 'lift'. It throws off a 'starting' vortex that is essential to the production of lift: it generates another when it is stopped. If it is stopped soon after the start, these two vortices will sometimes be seen to collide and annihilate each other because they rotate in opposite directions. The shadows of the flow will show a wake in which the water to be deflected sideways thereby producing 'lift'. Experiment with different angles and speed of motion.

Vortices in gases differ from those described above because, although both forced and free gas vortices exist, there is no 'free' surface of uniform (atmospheric)

pressure as with liquids. However, the circular trajectory of the gas particles necessitates a radial pressure gradient with lowest pressure at the centre. One frequently observes the wind induced rotational motion of dead leaves, or waste paper, when stuck in a convex corner of a building. Aircraft wings generate 'trailing vortices'. These arise from the difference of pressure below and above the region of the wing near the wingtip. The pressure and temperature fall at the centre of these vortices and if the ambient air is cold enough, water condenses out. Because of their property of persistence explained above, the trailing vortices of large aircraft have been known to destabilise smaller aircraft following them in to a landing strip; this is why minimum separation distances are strictly enforced. Fig. 3.18(a) reveals the wing tip vortices generated by an aircraft leaving the upper surface of the cloud. The 'channel' formed in the cloud by the passage of the aircraft clearly demonstrates that the aircraft displaces air downwards in the process of producing lift. Fig. 3.18(b) shows a cross section through the trailing vortices shed by a heavy transport plane. The cover of this book shows a tip vortex generated by a crop duster aircraft.

Fig. 3.18(a) The trailing vortices that drive down the central trough are clearly seen 'written' in the cloud top

The trailing vortices contain substantial kinetic energy and contribute significantly to aircraft drag. This component is termed 'induced' drag: the vortices induce an increase in the effective angle of attack of the outer wing sections. The wings of many modern aircraft, including gliders, are fitted with tipped-up winglets which modify the flow around the wing tip and are said to reduce induced drag. They actually counter drag by exploiting the vortical tip flow to increase thrust.

Fig. 3.18(b) Smoke reveals the trailing vortices of a heavy transport aircraft

    If one follows a car with a rather smoothly curved top in misty weather, one may observe vortical flows trailing behind the roof. These are produced by the pressure gradient between the rear side surface and roof of the car which produces an upward flow. The front wheels of bicycles carried sideways on the rear of cars are sometimes seen to rotate as they protrude into these vortices. Air vortices are vital to the generation of lift by the flapping wings of many insects and also feature in the flow over the wings of some birds. These and other influences on the flight of aircraft and living creatures are described later in this chapter.

    The vortices generated by the edge of solid fences and other non-streamlined obstacles have been mentioned earlier. If a cylindrical structure is exposed to cross flow, vortices are generated periodically as the boundary layers on opposite sides leave the rear surface, as shown by Fig. 3.19. This creates an oscillatory side force on the cylinder which causes vibration that can be destructive. The phenomenon can be suppressed by attaching helical strips (strakes), as often seen on industrial chimney stacks. The tonal whistle made by cables in wind is caused in the same way: it does not depend upon cable vibration. The frequency of the tone is proportional to the wind speed and inversely proportional to the cylinder diameter. The phenomenon may be easily demonstrated by whipping rods of various diameters though the air. The sounds are not of a single frequency because the speed of movement increases from hand to tip.

Fig. 3.19 A vortex 'street' generated by cross flow over a cylinder (flow from left to right)

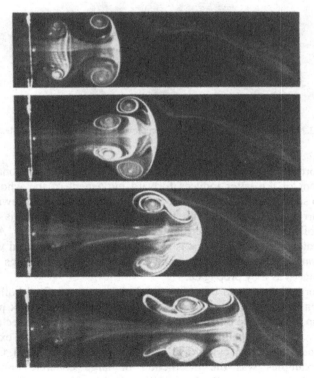

Fig. 3.20 Dance of the smoke rings

Fig. 3.21 Destruction of Ferrybridge cooling towers by wind-generated vortex shedding

Vortices may take the form of lines or rings. The rotational flow field of each segment of the latter induces forward movement of the whole structure, as observed in smoke rings which persist and move far beyond the blowing range of the smoker, carrying the smoke with them. The vortical rotation is caused by the separation of the air moving over the teeth and lips of the smoker. A pair of smoke rings generated a short time apart is shown in Fig. 3.20. The fluid velocity field of each acts on itself and also its partner, which induces the rings to periodically exchange places as they propagate.

Vortical flows can be put to good use, as in gas-fired boilers and bagless vacuum cleaners. In the latter, the low pressure core of the vortex produced by the fan sucks up dust and the circular flow ejects it radially into a collector. Vortices can damage or destroy pipelines, power cables, cross-flow heat exchanger tubes, chimney stacks and other cylindrical objects exposed to the wind. One of the most dramatic examples was the destruction of large concrete cooling towers at Ferrybridge in the North of England as illustrated by Fig. 3.21. Vortices in the form of tornadoes, in which the core comprises a free vortex with extremely high wind speeds and very low pressure, can also be highly destructive. The nature and characteristics of large-scale atmospheric vortices such as tornadoes are described in Chapter 5.

3.10   FLUTTER

You will have seen long thin leaves, such as those of field irises and reeds, vibrating and twisting vigorously in a light wind. This phenomenon is called 'flutter'; it involves the effect of the vibration of a structure on the aerodynamic forces acting on that structure, and *vice versa*. Flutter has been responsible for the crashes of many prototype aircraft in the past, although modern methods of analysis and test have largely eliminated it, but not completely. In 1992, Taiwan's IDF fighter project was shelved after the tailplane (stabiliser) of the prototype fluttered and the aircraft crashed.

All aircraft wings can vibrate up and down and twist at certain fixed natural frequencies. Positive vibrational twist (leading edge up, trailing edge down) increases the angle of incidence, and therefore the lift. Negative twist decreases the lift. In some forms of wing design, the maximum positive twist (and therefore lift) occurs as the wing moves upwards through its undisturbed (non-vibrating) position with maximum vibration velocity. Force times velocity is power, so the air injects vibrational energy into the wing. When the wing reaches its maximum upward displacement the twist is zero. As the wing oscillates down again at maximum negative velocity through the static position, the twist has by then reversed and therefore decreases the aerofoil lift. This again injects vibrational energy into the structure. If the rate of injection of vibrational energy exceeds the rate of dissipation of energy into heat by the damping of the structure, positive feedback occurs and the vibration amplitude will increase with time until the structure fails. [You can simulate the motion, although not the associated aerodynamic force, with your outstretched palm. Begin with your palm facing the floor, horizontal and held at a height lower than the shoulder. Now slowly move the hand upwards simultaneously giving it an increasing upwards speed and an increasing amount of positive twist. As the hand moves above the shoulder height begin to decrease the upward speed and twist until the latter is zero at maximum upward displacement. Reverse the process on the way down. You will observe that the positive lift combines with upward motion and the downward 'lift' combines with downward motion, the air doing work on the wing during both stages.]

It is not only aircraft wings that flutter; tail fins and tailplanes can also suffer this form of instability. Flags exhibit an extreme form of flutter which is much more complicated than wing flutter. Bridges have been known to flutter. In 1940, the deck of the Tacoma Narrows suspension bridge in the US State of Washington oscillated simultaneously in vertical and twisting motion, fluttered and collapsed into the river. The wind speed was about 42 mph. No lives were lost, although cars were on the bridge at the time.

3.11   SUPERSONICS AND SHOCK WAVES

Everyone is familiar with the bow waves produced by a boat. However, distinct bow waves will not be produced if the boat travels sufficiently slowly. In this case, a set of weak, almost circular ripples will spread out all around the boat. At

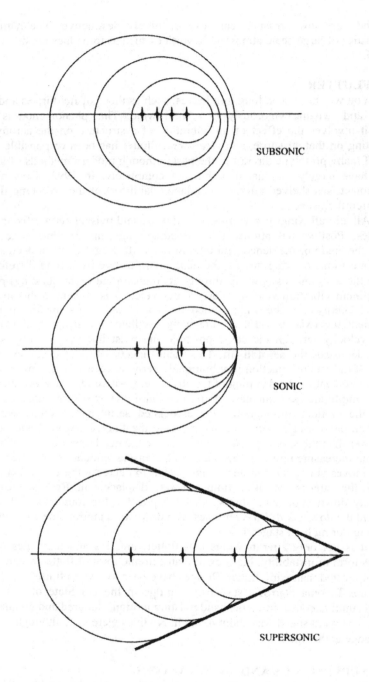

SUBSONIC

SONIC

SUPERSONIC

Fig. 3.22 Disturbances pile up into a shock wave
at supersonic speeds

some speed, the boat will move faster than the ripples and no disturbance can then proceed in advance of the boat. As the boat's speed increases further, the successive ripples will combine to form distinct and strong straight crested bow waves at an increasingly acute angle to the path of the boat. A similar wave will radiate from the stern where the water closes behind the boat. Weaker waves will be seen to radiate from small irregularities in the waterline profile of the hull. The energy carried away from the boat by these waves is the source of wave drag that can be minimised by hull design, such as that seen in 'fast' catamaran ferries.

The bow wave phenomenon exhibits features that are similar to shock waves that radiate from an object travelling through the air at a speed close to, or greater than, that of sound (supersonically). The 'message' that the air has been locally disturbed, say by a butterfly flapping its wings, is passed on, via inter-molecular 'collisions', to the surrounding air. Clearly, the speed of transmission of the disturbance is limited by the average speed of movement of the molecules in any one direction. As we shall see in Chapter 4, this is the speed of sound, which in air is about 343 metres per second at 20° Celsius. Provided that bodies move through the air at a much lower speed, the message to molecules ahead to 'get out of the way' can be successfully passed on in good time. However, if an object moves through the air at a greater speed, the message does not arrive in time. There must be a 'pile up' of molecules along a kind of 'bow wave' which is called a 'shock wave'. This is extremely thin close to the source (order of micrometres) and involves substantial conversion of flow energy into internal energy (heat). Because the temperature is higher than the ambient air, it propagates at a speed greater than that of sound in the ambient air. This process is responsible for the large increase in drag experienced by a moving body as it approaches and exceeds the speed of sound. This is the origin of the term 'sound barrier' - which is, of course, not a true barrier. The ratio of speed of motion to the local speed of sound is defined as the 'Mach Number', denoted by M and named after an Austrian physicist. The transition from subsonic (M<1) to supersonic (M>1) speed is simplistically illustrated by Fig. 3.22. The 'bow' and 'tail' shock waves make an increasingly acute angle to the flight path as the body's speed increases.

Because the *local* airflow speed relative to a moving body such as an aircraft or bullet can exceed the relative speed of the body itself, as on the upper surface of a lifting wing, shock waves appear at such locations at a Mach number less than unity based upon body speed. This is illustrated by Fig. 3.23: note the shock sitting over the canopy. The shock waves generated by a supersonic aircraft are heard as 'sonic bangs' (or 'sonic booms') on the ground and the associated air movements have been known to drive flocks of birds into the sea. Current aeronautical research aimed at eventually allowing commercial supersonic flight over land is testing the idea of initiating weaker, less concentrated shock waves by mounting a long, thin probe on the nose of aircraft and another at the rear, with the aim of spreading out the shocks as they hit the ground. This would make their sound less objectionable to human beings and less frightening to animals, but could increase their potential for causing damage to structures such as glasshouses.

Fig. 3.23 Localised shock waves on an aircraft flying just below Mach 1

3.12   BEND IT LIKE BECKHAM

An aerodynamically generated phenomenon that is influential in almost all sports involving the flights of balls is that of swerve induced by ball rotation. Tennis players, baseball pitchers and footballers (except goalkeepers) love it: golfers hate it. From the earlier explanation of aerofoil lift, we know that sideways force is associated with pressure difference between the two sides of a moving body and a sideways deflection of the flow in the opposite direction to the force. Rotation causes the surface of one side of the ball to move in the same direction as the centre of the ball, thereby increasing its speed relative to the air. The effect is to cause early boundary layer separation which inhibits the external flow from following the curve of the ball's surface and from producing a substantial reduction of surface pressure. On the other side, rotation reduces the speed of the surface relative to the air, sometimes to zero. The boundary layer stays laminar, thin and attached to the surface over a greater arc than on the opposite side, allowing the external flow to curl further round the ball, leading to reduced surface pressure, an imbalance of pressure forces and a net sideways force. The flow is illustrated by Fig. 3.24 in which the angular deflection of the flow is clearly seen. The effect is strongest over a fairly small range of ball speed that depends upon the ball diameter.

A similar mechanism operates when air flows over rotating cylinder. This is known as the 'Magnus effect' after a German professor who explained it in 1854. [In fact, the effect had been discovered and qualitatively explained by the English artillerist Benjamin Robins some 110 years before, in relation to the curved trajectories of rotating shells.] The cardboard tube and thread may be used to demonstrate this effect. With the tube hanging from its threads, rotate it so as to wind the cotton up to the top and ensure that the tube is horizontal before release. Ask a friend to observe the trajectory of the tube after you release it. This is a simpler flow system than that over a ball because the boundary layer flow on a ball is three-dimensional. Measured flow patterns around a cylinder rotating at a range of speeds are shown in Fig. 3.25. The parameter u/v is the ratio of the speed of the surface due to rotation to the speed of the oncoming flow. The asymmetry of the flow, the curvature of the streamlines over the upper surface, and the difference between the directions of the flow upstream and downstream of the cylinder clearly indicate that a lift force is generated that increases with rotational speed.

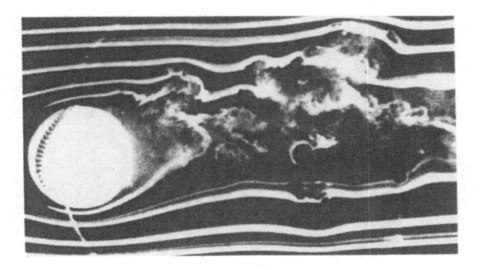

Fig. 3.24 Flow around a baseball rotating in an anti-clockwise direction

The effect was exploited by Anton Flettner who, in 1922, submitted a patent application for his 'rotor ship'. Rotation of a funnel-like cylinder produced a horizontal force in a wind. A large two-rotor ship named 'Buckhau', seen in Fig. 3.26, sailed successfully from Danzig (now Gdańsk) to Scotland and back in 1925 and to New York via South America in 1926. Sadly, the inability of the rotor ship to make way under windless conditions, together with the availability of cheap fuels for conventionally powered ships, sealed its fate.

Fig. 3.25(a) u/v = 2

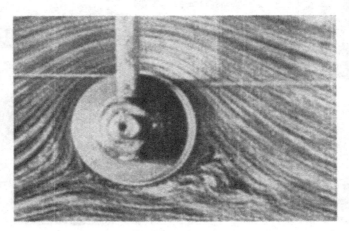

Fig. 3.25(b) u/v = 3

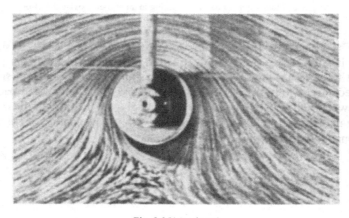

Fig. 3.25(c) u/v = 4

Fig. 3.25 Flows from left to right around a clockwise-rotating cylinder

Fig. 3.26 The Flettner-Rotor Ship

Cricket bowlers exploit another related mechanism to 'swing' the ball in an attempt to defeat the batsman. A cricket ball has one, raised, continuous, circular seam, unlike the seams of a baseball. Upon delivery, the bowler keeps the plane of the seam upright, but angled at about 20° to the 'flight' direction. On the side of the ball where the seam is closest to the front of the ball, it destabilises (trips) the boundary layer to make it turbulent. As explained earlier, in connection with the golf ball, this more energetic form of boundary layer 'sticks' more effectively to the ball's surface to delay separation, allowing the air to curl around the ball to produce low surface pressure. By contrast the laminar boundary layer on the other side separates easily and the airflow curves much less so the pressure is higher. This combination causes deflection of the wake from the direction of travel of the ball, as shown in Fig. 3.27, which betrays the action of a sideforce.

Bowlers polish the 'laminar' side of a ball to suppress boundary layer transition, which enhances the effect. The skilful swing bowler can ensure that the ball does not rotate during its passage, because rotation destroys the swing mechanism. One of the features of the swerve and swing mechanisms that skilful players exploit is that they operate most effectively over a limited speed range. This explains the observation that cricket balls can swing more strongly after they have bounced, lost speed and passed the batsman. Interestingly, when the whole surface of the ball becomes scuffed and rougher, the ball can swing in the other direction; this is known as 'reverse swing'. It seems that the roughness induces the boundary layer on the formerly smoother side of the ball to turn turbulent earlier than on the seam side: but the jury is out on this explanation. Strangely, no scientific evidence has been produced to support the contention that a cricket ball swings more readily under cloud cover - but players swear that it does. The 'meandering' form of the seam on baseballs produces a complex swerving effect in which spin causes deviations that vary with time and direction during flight.

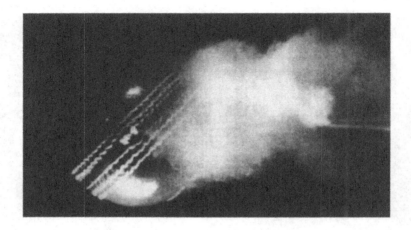

Fig. 3.27 The wake of a cricket ball travelling from right to left

3.13   SOME OTHER FLIGHT MECHANISMS

Previous sections explain the basic mechanisms of the generation of lift and drag by the aerofoils of conventional fixed wing aircraft. But there are other forms of aircraft that employ different configurations and flight strategies; and there are also many non-aeronautical examples of flight by both man-made systems and natural creatures. The following examples are briefly presented to show how air can be used to sustain flight by a range of different mechanisms, some of which are still not fully understood.

Flow separation from the upper surface of conventional wings occurs if the angle of incidence to the flight direction exceeds about ten degrees. This is bad news because it seriously reduces lift and produces 'stall' which is a potentially dangerous condition. This problem is overcome when landing and taking off by deploying movable mini-wings on the leading and trailing edges of wings, but they also greatly increase drag and are therefore unacceptable for efficient flight. If the leading edge of a wing is swept back by a sufficient amount, to form a 'delta' configuration, and the leading edge is sharp, the flow regime over the upper surface is quite different from that over conventional wings. In a manner similar to the generation of the trailing vortex from the tip of a conventional wing, flow separates from the forward region of the leading edge to form a 'leading edge vortex' (LEV) which travels back over the surface of the outer part of the wing. As we have seen, the pressure in the core of a vortex is low because of flow rotation. This produces a large amount of lift. Fig. 3.28 shows the LEVs on the leading edge extensions of a supersonic fighter aircraft. Notice the condensation caused by the low pressure and temperature in the core of the vortex. The Anglo-French airliner Concorde could land and take off at high angles of incidence without needing ancillary devices because the LEVs were present and stable over a wide range of angles of incidence.

Fig. 3.28 Leading edge vortices (LEVs)

Aerofoils produce lift when moved through the air and an alternative to linear translation employed by fixed wing aircraft is to rotate blades of aerofoil cross-section. This is the principle of helicopter flight. By tilting a rotor it can be made to produce both lift and thrust. Another rotor in the form of a propeller is mounted in a vertical plane at the end of the helicopter tail to prevent the main rotor from rotating the whole aircraft. A less common form of rotorcraft, called an autogyro, has a freely rotating set of aerofoil shaped blades above it and a propeller to provide forward thrust; the forward motion rotates the blades to produce lift.

Rigid wing gliders, or sailplanes, employ wings having conventional aerofoil sections. To stay aloft for long periods the pilots seek out regions of rising air called 'thermals' in which they circle. A paraglider is a flexible canopy which has a flexible top surface sheet and a similar bottom sheet that are connected together by a set of aerofoil-shaped interior ribs that form a set of cells with leading edge openings. The stagnation pressure in forward flight keeps the cells open and the canopy taut, to provide a lifting surface. Sky diving canopies are similar in function. Many birds, such as the albatross, eagle and buzzard stay aloft for long times and can fly considerable distances with hardly a movement of their wings. In gliding and soaring they exploit updrafts of the air to provide the angle of incidence and lift that counteracts gravity. Unlike rigid gliders they continually adjust the shapes of their wings to maximise the lift effectiveness. Gliding swifts have sharply swept-back wings and generate LEVs on the outboard sections, as shown diagrammatically in Fig. 3.29.

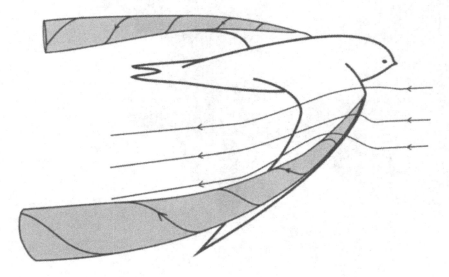

Fig. 3.29 LEVs over the wings of a swift (diagrammatic)

Another basic flight strategy is wing flapping. It is obviously no good to simply flap wings up and down, as seen in old films of some early 'ornithopters' (bird-like flight vehicles), because any lift produced by pushing air down is almost completely cancelled by the upward flap. (However, the flapping mechanism did act as an excellent vibration generator which often destroyed the vehicle.) So how do the birds and bees do it? There is no simple answer because the motions of the wings and the associated airflows are diverse, complex and not fully understood. A few examples are selected as a taster. Not so long ago, the flight of heavy compact insects such as bumble bees was a mystery, because scientists calculated the lift by assuming that the wings acted as simple fixed aerofoils: the answer was about three times too low to support the weight of a bumble bee. In recent years, techniques of visualising the flow over insect wings in wind tunnels have been developed. These reveal that insect wings do not simply flap up and down rigidly, but change their shape while moving. The generation of LEVs by the down stroke is the principal mechanism of insect lift. Stable LEVs have been observed on the upper surfaces of the wings of butterflies and dragonflies, among others. The amazing aerobatics of house flies, involving very large accelerations, are enabled by rotation of the wing plane through a large angle to give lift in both up and down flaps. Bees and hoverflies can also do this. In taking off, some insects, including butterflies, perform a 'clap and fling' procedure in which the wings are clapped together and then progressively separated in such a manner that vortices are generated as air flows in between them. Interaction of wings with their own wakes may also contribute to lift. Readers interested in butterfly flight are recommended to go to http://users.ox.ac.uk/~zool0206/s&t02.pdf where some wonderful images of vortices can be seen. The flapping frequency of insects varies widely: honey bees flap at about 200 Hz, but one variety of small midge has been recorded at over 1 kHz..

   Bird wings are generally not so flexible as insect wings and they principally

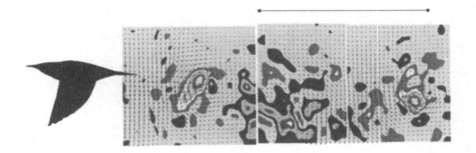

Fig. 3.30 Flow velocitiy vectors of vortices shed by flapping wings: flight is from right to left

use the down stroke to generate lift and thrust. One mechanism for avoiding lift 'cancellation' consists in separating the outer feathers and drawing the wings closer to the body in the upstroke, in much the same way as a breast stroke swimmer does on the completion of the thrust stroke. Some birds rotate the wing to provide lift also during the upstroke. This technique is employed by hoverers such as humming birds to enable hover without wind assistance, unlike hawks. Pigeons in slow flight through a smoke cloud were observed to generate smoke rings during each downstroke as early as 1938. The arrows in Fig. 3.30 indicate the velocity vector distribution around vortices generated by the downstrokes (two) and the upstroke (one) of a thrush nightingale flying freely in a wind tunnel. During gliding and soaring by large birds, such as albatrosses and giant petrels, steady state aerodynamics is the main principle used. Albatrosses have a rounded leading edge even on the hand wing (outer section) and can hardly use anything else. In common with aircraft, some birds deploy auxiliary flow control surfaces to suppress flow separation at high angles of wing incidence in slow forward flight, for example during landing, as illustrated by the use of the 'alula' (or 'bastard wing') by the Asian Openbill Stork in Fig. 3.31. However, understanding of all the mechanisms of bird flight is still incomplete.

Fig. 3.31 Open-beak stork deploying the lift-enhancing alula

Birds and insects can fold and rotate their wings during flight, but bats have many other options because their wings comprise membranes stretched over ribs that allow them to alter the whole wing configuration to adapt to different flight manoeuvres (Fig. 3.32). In simple forward flight the wing is fully extended and given considerable camber to increase the lift. The flapping downstroke generates a vortex that stays attached to the wingtip. During the flapping upstroke, bats fold their wings much closer to their bodies than birds thereby minimising the associated energy consumption. The versatility of wing shape allows a bat to make a 180-degree turn in a distance of less than half a wingspan.

Fig. 3.32 Bats' flight manoeuvres

Hovercraft do not truly 'fly' because they derive their lift by transferring their weight to the ground or water through an 'air cushion' which is pressurized by forcing air that is injected into the cavity below to take an outwardly curved path as it escapes under a flexible peripheral skirt, as shown in Fig. 3.33 This is an excellent example of the fact that it takes a transverse pressure gradient to generate a curved flow of air.

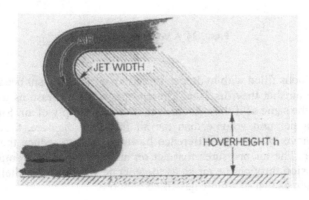

Fig. 3.33 Flow under a hovercraft showing how the lifting pressure is generated by forcing air to follow a curved path

Wing-in-Ground effect vehicles (WIG), sometimes referred to as a 'Flarecraft', exploit the fact that the lift generated by a wing at a given angle of incidence is considerably increased by the close proximity of a land or sea surface beneath it. One example is shown in Fig. 3.34. The concept of ground-effect marine vehicles was pioneered in the Soviet Union. The most famous example is the enormous, 500 ton military transport vehicle called the Ekranoplan, shown in Fig. 3.35. Development was halted by the precarious economic situation of the Soviet Union prior to the fall of the 'Iron Curtain'. Outside the Soviet Union, the most advanced developments of WIGs took place in Germany. These have led to the production and operation of small commercial vehicles. One of the disadvantages of the WIG is that it needs extra propulsive power to get moving through the water until wing lift becomes fully effective. One possible solution is to combine the WIG with a hovercraft-type system.

Figs. 3.34 A Wing-in-Ground vehicle

Balloons filled with hydrogen or helium air are buoyant because they weigh less than the air that they displace. The air atmospheric pressures on their external surfaces are the same as if they themselves consisted simply of air. Since they have a lower weight per unit volume than the air that they displace, they float. Hot air balloons achieve this weight difference by using heat to reduce the density of the contained air. The air pressures that act on airships are slightly modified by their slow progression through the air so as to generate weak lift, but their buoyancy has the same origin as that of simple gas-filled balloons.

Fig. 3.35 An Ekranoplan

Here are some interesting aerodynamic questions that you might like to consider. Many readers will have noticed cup anemometers rotating at the top of posts that carry meteorological measurement instruments. These cup anemometers, that measure wind speed, consist of a rotor in the form of a cruciform set of spokes that is free to rotate about a vertical axis; at the end of each spoke is attached an empty hemi-spherical cup. The open faces of the cups face in a tangential direction to the circle on which they lie and 'follow' each other around a circular path. The assembly rotates at a speed that depends upon the wind speed. The first question is as follows. Which of the three following aerodynamic forces do you think produces the greatest torque on the rotor? (i) The forces on the inner concave surfaces of the cups due to the wind blowing directly into the cups; (ii) The forces on the outer convex surfaces (backs) of the cups due to wind blowing directly onto the backs; (iii) The forces on the cups due to the wind blowing tangentially across the faces of the cups.

And, what effect does the speed of a cup have on the aerodynamic forces on it, given that its speed depends upon the wind speed?

I don't know the answer. I look forward to receiving your answers at frank.fahy@gmail.com Googlers should own up!

4

Sound in Air

Be not afeard: the isle is full of noises,
Sounds and sweet airs, that give delight, and hurt not.

William Shakespeare ' The Tempest' 1611

4.1   WHAT IS SOUND?

Sound in air is both a physical phenomenon and a subjective perception - we both produce and hear sound. The scientific term for the science of sound is 'acoustics' which derives from the Greek word 'akoustikos' which means 'pertaining to hearing'. This chapter concentrates upon the physical nature of sound in air, the effect of air state on its propagation, and the mechanisms of sound generation. The last aspect is technically complicated and difficult to explain without using mathematical models; but an attempt will be made in purely physical, qualitative terms.

Sound is ubiquitous. It is almost impossible to escape from it. It is a physical phenomenon that plays a vital role in human activity and society. It is one of the principal means of communication between humans, between other animals, and between humans and domesticated animals. It provides us with information about our environment. It carries signals from long distances, from every direction, and from unseen sources. It can be sensed over an enormous range of level and frequency by our auditory (hearing) system which is exquisitely sensitive. The frequency range of normal hearing of young persons is from about 20 hertz (Hz or cycles per second) to about 16,000 Hz. 20 Hz -20kHz is called the 'audio frequency' range. The fundamental frequencies of the notes of a modern upright piano (which correspond to their pitches) range from about 30 Hz to 4000 Hz; but the overtones of the notes, which govern their 'timbre' (sound quality), extend to much higher frequencies. The hearing range naturally decreases with advancing age and can be severely, and irreversibly, compromised by regular and extended exposure to excessive sound levels. The sense of hearing is at least as important, and possibly more important, than the sense of sight. It informs us about activities which are out of sight. Our sound sensing system is on guard even during sleep, when sight is blind. And sound affords us the universally valued pleasure of music.

We all know that sound travels in the form of waves but, to many readers, waves in the air may seem somewhat mysterious. The common feature of all mechanical (non-electromagnetic) waves is that they involve two interacting physical properties of the media through which they travel. One is the property of mass (or inertia); the other is the resistance of elements of the supporting medium to deformation. We most commonly associate the word 'wave' with waves on the sea or lakes. Water clearly has mass. If the level of its surface is disturbed locally so as to form a 'hump' or a 'dip', the local surface slopes from the horizontal. As we have seen in the previous chapter in relation to a vortex in stirred liquid in a cup, the associated hydrostatic pressure at any point below the surface is produced by the weight of the column of water above it. If the surface slopes, the pressure on any *horizontal* surface below the surface varies with distance along the direction of the slope; and we have also seen in the previous chapter, a pressure gradient produces acceleration of the fluid. This constitutes the mechanism that 'attempts' to restore the equilibrium (undistorted) state. In this case, the restoring force is due to gravity. In deep water, the fluid particles involved in a wave execute circular paths in the vertical plane, returning periodically to their original positions. As experienced by a rower in a small boat on the sea at a good distance from the shore (or rocks), there is *no bulk movement* of the water in any direction: it does not flow. What does 'flow' is the disturbance from the static state together with the associated kinetic and potential energies. Water waves clearly transport energy as evidenced by current attempts to harness it to generate electricity, and the wave damage to sea defences, among many other effects. Propagation of a disturbance without net directional flow of the supporting medium is a universal feature of mechanical waves, although in some extreme examples such as explosively generated waves, and water waves breaking onto the shore, bulk flow also occurs.

Air clearly has mass; the atmospheric pressure to which we are subjected is due to the weight of the mass of the air above us. Like all fluids, air is compressible and resists any attempt to change its density. Air at sea level is about 15,000 times as compressible as water. If you put your finger firmly over the outlet of a bicycle pump and attempt to push in the plunger, you will feel its resistance which acts like the stiffness of a mechanical spring: the shorter the pump, the greater the air stiffness. This is a manifestation of the rise in temperature, and hence, of molecular speed, generated by the work done on the fluid by the inward movement of the applied force, together with the increased rate of impact of the air molecules on the piston and, hence, of pressure. The interaction between the mass of air and its stiffness is the essential mechanism of acoustic wave motion in air and other gases.

Before dealing with the propagation of sound waves we shall consider a simple acoustic system which can be understood without any explicit reference to waves. This is the acoustic resonator, named after the great German physicist Hermann von Helmholtz. It consists of a vessel that has a distinct neck that connects the contained air to that outside, as in Fig. 4.1. The air in the vessel acts as an elastic spring and the air in the neck possesses mass. Just like a ball hanging from a spring, the system has a *natural frequency* at which the mass *freely* vibrates up and down after being given an initial displacement and released. A large amplitude response

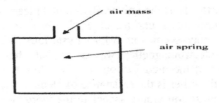

Fig. 4.1 Simple Helmholtz resonator

termed '*resonance*' occurs if the system is periodically disturbed at this frequency. The Helmholtz resonator may be demonstrated by blowing *gently across* the opening of the neck of a bottle having a distinct neck, which will elicit a musical tone. [Beer bottles are very popular for this purpose because one has to empty them before the experiment.] The stiffness of the air in the vessel is inversely proportional to its volume. This can be demonstrated by partially filling the bottle with water; the resonance frequency, which is proportional to the square root of the stiffness, increases. Helmholtz invented acoustic filters in the form of a set of spherical vessels of various sizes that had open necks on one side and very small pierced 'teats' that fitted into the opening of the ear canal on the other. Each one was tuned to a different resonance frequency so that external sound was preferentially filtered before entering the ear. Lacking modern electronic frequency analysers (that were not to be invented for a further 120 years), Helmholtz used this cunning technique to investigate auditory frequency perception and sensitivity.

The stiffness of air within a shallow cavity, such as that between double walls in buildings, is inversely proportional to the cavity width. This has important implications for the design of lightweight double walls to minimise the transmission of sound. To provide adequate low frequency sound insulation the panels of such walls should be separated by at least 150mm because the masses of the panels resonate on the stiffness of the air in the cavity to seriously degrade the sound insulation. The cavities should contain sheets or slabs of sound absorbing material such as mineral wool to suppress other, higher frequency, acoustic resonances of the cavity. Panels of double glazing are typically 19 mm apart. A greater separation would allow convection currents of air to flow within the cavity which degrade the heat insulation. But, the stiffness of the air in the cavity is such that the panels are effectively dynamically locked together over much of the low frequency range, which degrades their sound insulation performance. This problem is minimised in the windows of recording and broadcasting studios by separating the double panels by at least 250 mm and lining the edges of the cavity (the reveals) with sound absorbent material

It might be of interest to pianists to learn that the stiffness of the air between the soundboard of an upright piano and a wall in *close* proximity has two important effects. It couples the soundboard to the wall so that acoustic vibrations are readily transmitted into the wall whence they can travel easily via the building structure into neighbouring rooms or adjoining dwellings. It also acoustically damps the vibration of the soundboard which adversely affects the tone of the instrument.

The process of sound wave propagation may be understood by analogy with waves propagated along the helical flat spring toy known as a 'Slinky' [see www.wikipedia.org/wiki/slinky].   Stretch a Slinky out straight on a smooth flat surface and suddenly displace one end along its length while holding the other end fixed. A dark band of squeezed-together rings alternating with stretched-apart lighter bands will be seen to pass along the slinky and reflect off the fixed end. The movements of the coils mimic the movements of the air particles in a sound wave; they return to their rest positions after the wave has passed. The dark band is analogous to the local increase of air density and pressure in a sound wave and the light bands are analogous to regions where the air density and pressure are reduced. As explained in Chapter 3, spatial variations of pressure (pressure gradients) cause accelerations of the associated fluid. *Thus, the density and pressure variations are intimately linked to the oscillatory motion of the air, and vice versa.* This interaction forms the essential mechanism of sound waves. Unlike this one-dimensional Slinky analogy, sound propagates in all directions in the air, but not usually uniformly.

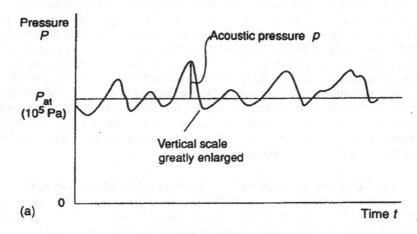

Fig. 4.2 Acoustic pressure (sound) signal

As with the deep-sea water wave, acoustic disturbances of the air at the sound levels typical of our everyday experience propagate without net flow of the air. The air particles simply vibrate about their rest positions. As they do, the air density and pressure fluctuate about their static (equilibrium) values by very small amounts; the magnitude of the pressure fluctuation is greatly exaggerated in Fig. 4.2. The maximum *fractional* variation in air pressure in a sound wave typical of the level of a fairly loud conversation as received by a listener at a distance of about one metre is about one part in one million. Sound pressures of which maximum fractional deviations are of the order of one hundred thousandth require the wearing of ear defenders to mitigate hearing damage. It is these very small variations of pressure that vibrate our eardrums and, via a complicated mechanical, fluid and neural path, produces the perception of sound by the brain.

The largest fractional variation of pressure that we can tolerate without severe discomfort is about two parts in ten thousand. The smallest variation at 1kHz that a healthy young person can hear, which is the 'threshold of hearing', is about two parts in ten thousand million. This is an operational range of one million which greatly exceeds the equivalent range of the eye. The magnitude of the oscillatory air displacement at 1 kHz at the threshold of hearing is six million times smaller than the diameter of an average human hair. The human auditory system is truly remarkable, and should be cherished and protected from excessive exposure to loud noise.

4.2   SOUND ENERGY

Sound waves transport energy in two forms: kinetic energy of the oscillating air particles and potential energy of fluid compression (as in a bicycle pump) and expansion. Fortunately, the rate of generation of sound energy in air by most common sources of sound is minute. *One watt* of sound power from a nearby source is sufficient to deafen you. The quoted power rating of conventional loudspeakers relates to the electrical power consumed, which is typically about 100 times the sound power radiated; the remainder is dissipated into heat in the voice coil yielding 1% electroacoustic efficiency. The total sound energy generated by a crowd during the entire duration of a football match in a large stadium is sufficient to cook only a couple of eggs. A male speaker engaged in normal conversation generates about fifty millionths of a watt. A large symphony orchestra playing fortissimo generates between 1 and 10 watts. A large airliner generates about 100 kilowatts of sound at take off. The very intense sound generated by aircraft jet engines has the potential to cause severe fatigue damage and subsequent failure so that engineers have to design them to resist this onslaught. The sound generated by the engines of rocket launchers, which radiate about ten million watts at launch, has been known to cause on-board electronic systems in multi-million dollar satellite payloads to fail, thereby wrecking the mission.

4.3   EFFECTS OF AIR PROPERTIES AND CONDITIONS ON SOUND

Acoustic disturbances in air at 20° C travel at a speed of 343 metres per second (m/s) = 767 miles per hour (mph). Sound speed in air at commonly experienced levels is *independent of frequency*; if it were not, we would be unable able to communicate by means of speech or music – think about it. For comparison, sound in water travels at the higher speed of about 1450 m/s because, although water is about eight hundred times as dense as sea level air, it is fifteen thousand times 'stiffer' (less compressible). The speed of sound in air is determined by the average speed of the molecular 'messengers' that pass on the acoustic disturbance from one to another: clearly, it can't travel faster. We have seen in Chapter 1 that gas temperature is proportional to the square of the average molecular speed. Thus, the speed of sound increases with square root of temperature. This has many important effects. We have seen in Chapter 3 that as the speed of an aircraft approaches the speed of sound (Mach One), its drag increases substantially due to the formation of shock waves. At ten thousand metres altitude the air temperature is about – 40° C = 233 K. This limits the economically viable speed to about M = 0.85 = 306 m/s = 684 mph.

This temperature dependence of sound speed has a very important effect on the propagation of sound near sea level. During most days, the temperature falls with height (lapse). This causes the direction of travel of sound waves to bend (refract) upwards as they move away from the source which causes a sound 'shadow', seen shaded in Fig. 4.3. This is beneficial in terms of the spread of noise from traffic and industrial plant. However, on sunny days following frosty nights, the air temperature rises faster than the ground temperature. In this case sound refracts towards the ground and this exacerbates noise nuisance. The condition where the air temperature

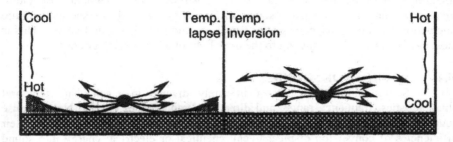

Fig. 4.3 Effects of a vertical air temperature gradient on sound propagation near the ground

rises with height (inversion) can also occur on summer nights when people like to sleep with windows open. An extreme example of refraction occurred during Saturn rocket launches, when people living within a few miles of the blast were exposed to less noise than those living thirty miles away.

Another form of refraction is caused by wind. As the air travels over the land it forms a thick boundary layer in which the air speed increases with height, as explained in Chapter 3. Sound is literally 'carried on the wind'; its speed relative to the ground is increased in the wind direction and decreased in the opposite direction. The associated refraction, illustrated by Fig. 4.4, causes a 'sound shadow' to be formed upwind of the source. Thus, it is more difficult to hear someone shouting from a downwind location than from an upwind location; a sound 'shadow' is formed, as illustrated by Fig. 4.4. It has nothing to do with the popular belief that sound is not able to propagate against a wind. Unlike light, which forms sharp shadows, sound bends rather easily around all except very large obstacles. One can talk to a neighbour over a garden wall, even if one cannot see them. This phenomenon is termed 'diffraction'.

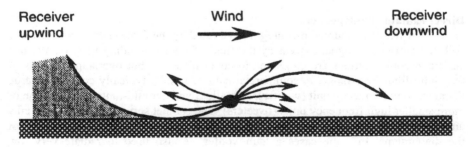

Fig. 4.4 Effects of a wind boundary layer on sound propagation near the ground

Air humidity has a significant effect on sound propagation over long distances. This is because water molecules can vibrate in more ways than those of oxygen and nitrogen. When the air molecules interact with the water molecules, the former transfer to the latter a proportion of their energies that is converted into forms of molecular vibration that do not involve the translational motion that we have previously associated with energy, temperature and pressure and sound propagation. This energy is effectively 'lost' to the sound wave. Consequently, the wave loses strength as it propagates: it is said to be 'absorbed'. The attenuation of audio-frequency sound increases as the square of frequency. Turbulence and temperature fluctuations in the atmosphere scatter (break up and redirect) sound waves. This is most easily observed by listening to the overflight of a high-flying aircraft.

4.4   SOURCES OF SOUND

Sound sources are almost infinitely diverse in mechanism, temporal characteristics, frequency range, and directional radiation of energy. Not all noise sources involve vibration of a solid body. They all generate sound energy, but their efficiencies of conversion of mechanical, chemical or electrical energy into sound energy vary widely. In spite of this diversity, it is possible to place sources into three basic categories, as illustrated by Fig. 4.5. The following examples are chosen purely to illustrate the basis of this categorisation. Strangely, very little, if any research appears to have been published on that very common sound source – the handclap. Please let me know if you are aware of any (frank.fahy@gmail.com).

4.4.1   Category 1 sources

Category 1 includes those sources by which *air volume is displaced* or *fluid mass is injected* in an *unsteady* fashion. They constitute the most efficient forms of source. It is actually the rate of change of the rate of volume displacement (volume acceleration) that generates the sound. This is illustrated by the sound of a handclap. Place your hands together close to your nose with your fingers pointing upwards and with the fingers of one hand resting on the palm of the other. Now clap your hands rapidly together. Your face will feel the increasingly strong outflow of air, but no sound will be heard until the hands collide. At this point the outflow rapidly *stops*, and the sudden change in the rate of air volume displacement (volume acceleration) 'stretches' the local air and slightly reduces its density; this is what causes the sound. The sound wave produced by the discharge of gas from an AK-47 rifle barrel is seen in Fig. 4.6, together with the shock wave generated by the supersonic bullet.

Direct radiator loudspeakers

A flexibly mounted diaphragm is vibrated by the force exerted on the voice coil that carries the signal current by the field of the surrounding magnet. Various sizes and configurations are employed to cover different, but overlapping, parts of the audio-frequency range. A commercial cabinet system typically contains a large diameter, low frequency unit (woofer), a smaller diameter mid-range unit and one or more smaller high frequency units (tweeters). The cabinet serves to support the units and also to prevent cancellation of displaced volume between the front and rear of the diaphragms. In some cases a 'sub-woofer' is also used to radiate very low frequency sound.

Fluctuating volume/mass displacement/injection

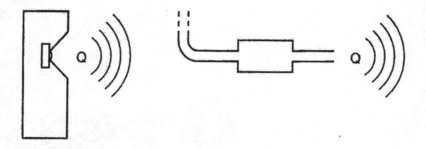

Accelerating/fluctuating force on fluid

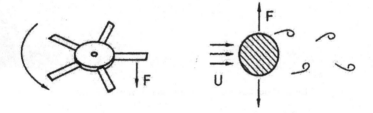

Fluctuating fluid shear stress

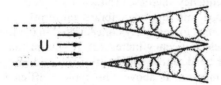

Fig. 4.5 Source categorisation

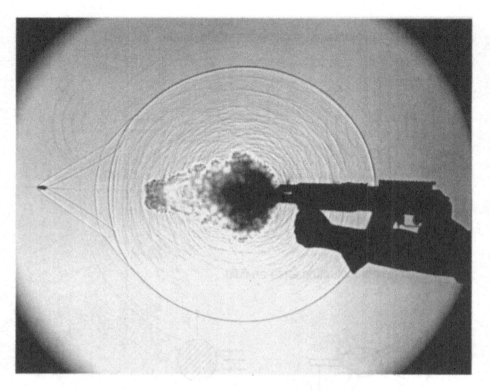

Fig. 4.6 Bullet shockwave and discharge sound wave of a rifle

The various sizes that cover different parts of the audio-frequency range are necessary for four principal reasons. The efficiency of sound radiation by a loudspeaker is low at frequencies such that the acoustic wavelength substantially exceeds the peripheral length of the diaphragm. (The wavelength in metres is 343/(frequency in Hz).) Hence, a large diaphragm is necessary to radiate low frequencies. But the mass of the diaphragm increases with size, and mass becomes increasingly 'difficult' to vibrate as frequency increases (Newton's Second Law of Motion). Large diaphragms 'break up' and do not vibrate uniformly at high frequencies. And the radiated field becomes increasingly concentrated in the forward arc as the acoustic wavelength decreases (frequency increases). If you mistune your radio to produce noise and move your head across the front of the mid-range unit you will experience this effect. As mentioned above, the typical efficiency of conversion of electrical energy to acoustic energy is between one and two percent. The rated power is the electrical power supplied: one watt of *acoustic* power would be deafening. The remaining 98 to 99% goes into heat in the electromagnetic driving mechanism. However, horn loudspeakers can reach an acoustic efficiency of up to 40%. An interesting characteristic of Category 1 sources is that if a pair of equal strength, operating in anti-phase (push-pull), are sufficiently close together, they largely cancel each other's radiation. Try switching the polarity of the connections to one of your stereo units and you will observe this effect, particularly on the bass frequencies.

Voiced speech, singing and whistling

The so-called 'voiced' sounds of speech are generated by vibration of the vocal cords which modulate (periodically vary in time) the airflow produced by the lungs. The temporal variation of the airflow rate is not sinusoidal, so the sound contains many equally spaced frequencies (harmonics) which gives it its characteristic quality. This is the 'engine room' of the voice. The vowel sounds are voiced. Say 'aaah'. Now say 'cat' slowly: the 'c' and the 't' are not generated by vocal cord vibration; nor is a whisper. How do you think these are generated? Hint: look at Category 2 below. Now sing 'aaaaah'; you can vary the pitch by tightening or slackening your vocal cords, keeping an unchanging configuration of your mouth and tongue. The tonal sounds of the singing voice are generated by the filtering effect of the acoustic resonances of the oral and nasal cavities on the multi-frequency sound produced by modulated flow through the vocal cords, in much the same way as Helmholtz's glass spheres filtered the passage of external sound into his ear. If you open your mouth wide and sing a note while maintaining a constant tension in your vocal cords, you can hear the effect of the shape of the oral cavity by pushing your lips forward to form an 'O'. Try whistling and slide the pitch up and down: feel what your tongue is doing to change the pitch. It's remarkable that such wet, softly lined, body cavities can resonate so sharply, as witnessed by the pure sound of a soprano singer.

Fig. 4.7 John Tyndall's Steam Syren

Sirens

Sirens are devices that periodically release compressed air or steam into the atmosphere. This is generally effected by means of opening and closing holes in a rotor that passes over a set of matching holes in a fixed plate. They are therefore similar in principle to the voice, but thousands of times more powerful. They were used in WW2 to give air raid warnings and to signal 'all clear'. They are mounted on pylons in many towns as danger warnings and are test sounded at fixed times of the week. In the nineteenth century, the famous Irish physicist, John Tyndall, employed an enormous steam-driven siren and horn, shown in Fig. 4.7, to investigate the propagation of sound through fog at the bequest of the Trinity House, the then lighthouse authority for England, Wales, the Channel Islands and Gibraltar. He also sent up explosive rockets for the same purpose.

Internal combustion engines

As the outlet valves of an internal combustion engines open and close, they produce a periodically varying flow through the exhaust pipe. This unsteady component of the flow out of the pipe radiates as a Category 1 source.

Champagne bottle

When one opens a bottle of champagne the gas escapes suddenly to produce the 'pop'.

Vibrating surfaces

Vibrating surfaces constitute very common sources of sound. For example, the sound that is transmitted through a wall is radiated by the vibration of the wall in response to the pressure field of the sound falling on its other side. Much of the noise of machines, including internal combustion engines, is generated by vibration of their surfaces. Although such sources do radiate by displacing air volume at their surfaces, the vibration fields are not uniform. Different areas vibrate with different amplitudes and phases; at any one instant of time, some parts are moving in one direction and others are moving in the opposite direction. If neighbouring areas vibrate in anti-phase (opposite directions), the radiation cancellation effect described above in relation to an incorrectly wired stereo system operates, particularly in the low frequency range. This feature makes vibrating surfaces impossible to categorise in a generic manner. We will simply state that if two sheets of the same size and material (e.g. steel), but of different thickness are *mechanically* vibrated at the *same frequency* with the *same amplitude*, the thicker one will radiate more strongly than the thinner one. But beware; this does not apply to sound transmission through walls! Heavier walls will almost always transmit less sound that lighter walls.

Tyre noise

The noise of road vehicle tyres, which dominates traffic noise at speeds above about 50 mph (80 kph), is generated by a number of mechanisms. Vibration of the tyre wall due to the unevenness of the road surface is one of the principal noise sources, but also important is so-called 'gas pumping'. As the tyre rotates, air contained in the tread wells is squeezed out when they contact the road and re-enters when they leave the road, thereby producing volume acceleration. The presence of water on the road greatly increases this noise due to ejection acceleration of water particles. Gas-pumping noise is reduced by the use of porous road surfaces which ease the entrapment and compression the air. The curve of the tyre amplifies this noise by acting as a form of acoustic horn. Slicks produce little gas-pumping noise, but have low skid resistance and poor water ejection and are not suitable for general use.

4.4.2   Category 2 sources
This category includes sources that radiate by applying fluctuating *forces* to the ambient fluid but which involve *zero net volume acceleration*. The general equation that governs sound generation in a fluid such as air includes a term that expresses the action of a force, in addition to one accounting for Category 1 volume acceleration. This category of source is much less efficient at radiating sound than Category 1 sources: examples follow.

Vibrating slender bodies
At frequencies where a vibrating tube or cable or pipe, or other slender body, has a diameter very much less than an acoustic wavelength, it produces almost no net volume displacement because, in simplistic terms, when the body vibrates transversely, the fluid displaced by the advancing half of the body 'slips' around towards the retreating half and cancels any net volume displacement. However, in order to generate the oscillatory acceleration of the fluid in 'sloshing' it to and fro, the body must exert an oscillatory force on it. A very thin structure such as a violin string radiates negligible sound itself. It vibrates the violin body, which is the principal sound radiator.

Turbulent flow acting on solid surfaces
Surprising though it may seem, the action of the unsteady action of turbulent flow in producing fluctuating pressures on a *solid* surface generates sound even if the surface does not vibrate. The presence of the surface constrains the fluid motion normal to the surface, producing density changes, and thereby a small proportion of the kinetic energy of the flow is converted into sound energy. You can demonstrate this for yourself by blowing closely on your finger tip or on the edge of a credit card. Blowing in the absence of the object produces much less sound which comes largely from the action of turbulence on your lips. This form of source accounts for much of the noise of wind blowing through trees and for the noise produced by the flow of air through the grilles or louvres at the exit of ventilation ducts. The sound power of Category 2 sources increases with the sixth power of the air speed; for example, doubling the flow speed increases the sound power by a factor of sixty four, or an increase of sound pressure level of 18 decibels, which is nearly four times as loud. This is why the speed of flow through the terminal devices of ventilation and air conditioning systems must be minimised, especially in auditoria (spaces designed for listening).
This category of source mechanism has recently become of great concern to engineers seeking to minimise the noise of landing aircraft because the noise of turbulent flows generated by devices deployed on the wings to increase lift and drag, together with that generated by the landing gear, termed 'airframe noise', is now comparable with engine noise. Environmental noise so generated is also a limiting factor on the speed of electric trains that sport overhead pantographs.

Vortex shedding

When air flows transversely over a structure of circular cross-section, such as a rod, it generates a sequence of vortices that are shed periodically into the wake, as illustrated by Fig. 3.18. This process generates an oscillatory force on the object, and associated sound, even if the structure does not vibrate. This can be demonstrated by swishing a rod rapidly through the air: the faster the motion, the higher the frequency. The 'singing' of wires and cables in the wind exemplifies this form of source. They do vibrate, but this movement generates little sound as explained in (i) above. Vortex shedding from cross-flow heat exchanger tubes of power stations has been known to 'cooperate' with acoustic resonances of the enclosed space to excite damaging vibration of the tubes. Other unwanted effects have been mentioned in Chapter 3.

Rotating aerofoils

Propeller, fan and turbine blades all act as aerofoils. They generate lift and drag like aircraft wings. In moving through a fluid, they deflect it to produce a predominantly steady flow and an associated thrust. The aerodynamic force on a blade is more or less constant, unless the inlet flow is made non-uniform by the presence of upstream flow obstructions, such as radial support struts and poorly designed duct bends. Sound is generated by rotating aerofoils, even in the absence of such obstructions, because the aerodynamic force rotates around the axis. The sound generated in this manner consists principally of a series of harmonics of the frequency with which the blades pass any point fixed in space (blade passage frequency) which increases with speed. Sound generation is greatly enhanced by upstream objects that partially block the inflow because the lift produced by a blade suddenly falls and then suddenly rises again as the blade passes into and out of the wake of the obstruction. You may hear the effect by running a desk cooling fan at top speed and placing an obstruction such as a ruler immediately upstream of the rotor. The tips of the propellers of some older forms of propeller driven aircraft, such as the Harvard, moved at supersonic speed. This caused extremely loud noise because of the generation of shock waves.

4.4.3   Category 3 sources

Category 3 sources produce neither *net volume acceleration* of a fluid nor *net force* on the fluid. They are extremely inefficient. A commonly experienced example is the 'clack' produced by the collision of snooker or pool balls and of children's marbles. The air remains in contact with them during impact and suffers acceleration and changes of density which radiate as sound. The sound is not made by vibration of the spheres because the lowest natural frequencies are above the limit of human hearing. Another such source is that of the struck tuning fork. The vibration of each tine constitutes a Category 2 source. But the tines vibrate in opposite directions and largely cancel the two individual sources. This is why the stalk is usually placed on a convenient flexible surface which it vibrates and enhances the sound. If the ear is placed close to the tines, and the fork is rotated about its axis, a figure-of-eight radiation pattern will be heard.

Another Category 3 source of far greater practical importance is that of jet flow. A jet of fluid issuing from a pipe mixes with the surrounding ambient fluid by means of viscous stresses. The shear layer at the interface between moving and ambient air is unstable and the resulting turbulence generates sound over a wide and continuous range of frequencies. The actual mechanism of sound generation is far too complex to be described in simple, qualitative terms. It was first explained just over 50 years ago by James Lighthill. Fortunately for the aircraft business, the efficiency of conversion of jet flow kinetic energy into sound is extremely low; although people regularly exposed to aircraft noise would be reluctant to accept this fact. For example, at take off, the engines of a large modern airliner deliver over one million newtons of thrust. The associated mechanical power is over one hundred million watts. A typical ratio of radiated sound power to mechanical power of the engines of large airliners in the mid-1990s was about 0.02%; this was reduced to about 0.003% by the installation of sound absorbent treatment in the air intake duct. By 2005, improvements in engine acoustic design and sound absorbent treatments had reduced this ratio to about 0.0004%. The associated sound power is about 400 watts. The sound power generated by a subsonic jet varies as the jet speed to the power eight; halving the speed reduces the sound power by a factor of 256 which translates into –24 decibels (a reduction of perceived loudness by a factor of over four). This extreme sensitivity has been exploited by the introduction of large by-pass turbofan engines in which the hot jet core is surrounded by a cold flow produced by a very large diameter fan, as illustrated by Fig. 4.8. Earlier, pure jet engines, such as those of Concorde, produced very fast, supersonic jet flows that were extremely noisy.

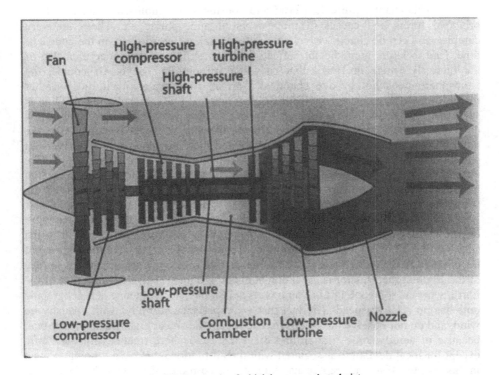

Fig. 4.8 Schematic of a high by-pass ratio turbojet

5

Meteorological Phenomena

The weather is always doing something there; always attending strictly to business; always getting up new designs and trying them on people to see how they will go

Mark Twain (Samuel Langhorne Clemens) 1876

5.1   ROTATION OF THE EARTH

This chapter presents a brief and considerably simplified overview of the principal forms and causes of movements of atmospheric air. These are very complex and can be chaotic, which means that small influences can, in the course of time, lead to large, unpredictable effects, as exemplified by the apocryphal flap of the butterfly wings in Brazil that causes a typhoon in Tokyo. Of course, the atmosphere does not behave chaotically on all time scales. If it did, weather forecasting would be fruitless. However, the complexity of atmospheric behaviour is such that it is very challenging to try to forecast the weather for more than a few days ahead (especially in a maritime climate like that of Great Britain). Here, we describe and explain the general mechanics and thermodynamics of air movement and briefly mention features of weather such as clouds and lightning. Before introducing the various generic forms of air movement and their causes we explain why observations of air movements made from surface of the Earth do not tell the whole story.

The Earth rotates about its axis at an angular speed of one quarter of a degree per minute. The associated speed of a point on the surface relative to the centre of the Earth is given by the product of the rotational speed and the distance of the point from the axis of rotation. This means that someone standing on the surface of the Earth at the Equator is travelling at about 470 m/s relative to the centre of the Earth, whereas someone in London travels at 293 m/s, and someone at a geographic pole has no relative speed. The fact that we experience still days with virtually no wind, and do not observe the clouds to rush across the sky at high speed, proves that, because of aerodynamic friction and drag, the atmosphere rotates with the Earth, except for local deviations associated with the air motions that we are going to meet in this chapter. If it, and the oceans, did not thus co-rotate, life on Earth as we know it would never have evolved.

5.2   FRAMES OF REFERENCE AND INERTIA FORCES

We now briefly digress to explain the problem of analysing the dynamics of the motion of the air relative to the Earth's surface; that is to say, motion as 'seen' by an Earthbound observer. Before specifically considering atmospheric motion, we must first consider the question of how motions of any material are precisely defined in scientific/mathematical terms. *It must be clearly understood that observed motion is not absolute: it is relative and depends upon the state of motion of the observer.* For example, one is sometimes fooled into thinking that one's (stationary) train is reversing into the station by seeing the forward motion of a train leaving the station on the adjacent track. The penny drops with the realisation that one feels no accelerating force acting on one's body. Suppose we were to specify the position of a lamppost in terms of the distance between it and a our chosen reference point fixed in a passing, *accelerating* car. We would deduce that the lamppost, together with the road, houses, etc., are accelerating in the opposite direction, which requires the application of substantial physical forces of which there is no physical evidence or cause. While sitting in a train taking a bend it is common to see objects sliding across a table and sometimes toppling over. From our perspective, sitting in our seat, it is as though some outward-directed 'pushing' force were acting on us and on the objects. In fact, no 'real' physical force in the common sense is acting in that direction. In fact, a 'real' physical force is acting on your bottom in the opposite direction to keep you going round the bend and not flying off like the loose objects. This is a manifestation of the property of matter termed 'inertia', which may be defined as a resistance to being accelerated, or equivalently, having its velocity changed in either magnitude (speed) or direction. (Remarkably, we don't really know why inertia exists!). But this definition begs the question - accelerated relative to what? Here we must pause to consider this fundamental question.

A 'frame of reference' in three-dimensional space comprises a set of three independent coordinates (spatial variables). The position of any point in space is specified by the values of these coordinates. The most commonly used frame of reference is the rectangular Cartesian coordinate system, of which an everyday example is represented by axes that lie along the three perpendicular edges of a room that meet in a corner. This coordinate system is named after the French mathematician and philosopher René Descartes who, it is said, while lying in bed one day, realised that the position of a fly in the room could be uniquely defined in terms of three distances measured along three perpendicular edges of the room. It might be thought that the surface of the Earth does not accelerate (except in earthquakes). However, as explained in Chapter 3, a point moving along a curved path is accelerating, even if its speed is constant: this is so-called 'centripetal' (Latin – 'centre seeking') acceleration. So, a frame of reference (or a 'stationary' observer) fixed in the surface of the Earth rotates with the Earth and accelerates towards its axis of rotation.

Newton's Laws of Motion (see Chapter 3) are valid only if expressed in terms of motions of material elements that are specified relative to so-called 'inertial frames of reference'; that is, ones that are not themselves subject to acceleration. But hold on! How can we assume that any frame that we choose is not subject to acceleration. Of course, we can't. We can only do the next best thing and choose a frame of which we are confident the acceleration is sufficiently small that any errors made in neglecting it will be negligible in terms of the behaviour of the system being modelled in terms of the application of Newton's laws. For example, when standing on the bathroom scales, we are accelerating towards the axis of the Earth (unless we are at one of the poles). The centripetal force necessary to accelerate us in this manner is equal to the difference between the 'downward' force of one's weight (gravitational attraction) and the 'upward' reaction force of the scales on one's feet, which must be less than the weight, since we accelerate. The latter force is what the scales register as our weight. For this reason, bathroom scales would register one's weight to be about 0.4 % less on the equator than at a pole. If the Earth rotated 250 times faster, the scales would register zero weight. An alternative way of explaining these effects is to postulate the existence of an 'upward' force acting in concert with that produced by the scales to balance the weight. This is known as 'centrifugal' (Latin –'centre-fleeing') force. It must be included in the equation of 'static' equilibrium of the person standing on the scales if the position of the person is measured relative to the local surface of the Earth (i.e., motionless) in order to compensate for neglect of the centripetal acceleration of that person..

Consider a stone stuck in the tread of a rolling tyre. Because the stone possesses mass (or, more generally, inertia) its 'adhesion' to the tyre, and consequent acceleration towards the axle, requires the action of a real, physical, centripetal force which is generated by friction between the tread and the stone. If the mass of the stone times the centripetal acceleration (the product equals the centrifugal force) exceeds the maximum available frictional force, the stone slips free and is no longer forced to accelerate; it flies off at a tangent with the same velocity (speed and direction) and kinetic energy that it possessed at the instant of release, subsequently moving as a ballistic missile under the action of gravity. It is released from centripetal acceleration by release of the centripetal force. *It is not, as often mistakenly imagined, 'thrown off' by centrifugal force.*

Consider another example. In order to stand 'still' on a rotating merry-go-round one has to cling on tight to one of the radial bars so that it applies a centripetal force to one's body. *In the frame of reference rotating with the merry-go-round* one is stationary and not accelerating and must therefore be subject to zero net radial force. The inwardly-directed force applied to one by the rail balances the outwardly directed inertia force, thereby keeping one fixed in position relative to the rotating frame of reference.

So we see that it is possible, and valid, to apply Newton's laws to the motion of material as expressed relative to a rotating frame of reference if we add terms to the dynamic equations to express the effect of components of acceleration associated with the rotation and the associated inertia forces.

5.3   THE CORIOLIS EFFECT

The first person to develop a rigorous mathematical analysis of the motion of *solid bodies* such as the components of machines, expressed in terms of a rotating frame of reference, was the French physicist Gaspard-Gustave Coriolis, in 1835. George Hadley, an English lawyer and amateur meteorologist, understood in the early 1700s that the trade winds that blow in the tropics in the northern hemisphere could only be explained if the rotation of the Earth were taken into account. However, his explanations of the causes were not comprehensive. A more complete understanding of the physics of atmospheric motion as influenced by Earth rotation was presented in the 1830s by the American, William Ferrel.

For practical reasons, it is conventional in dynamic meteorological models to define air movement *relative to rectangular Cartesian frames of reference fixed in the surface of the Earth*, having one coordinate axis perpendicular to the local surface of the Earth (vertically upwards), the other two lying on a plane tangential to the surface (on the floor). This procedure is sensible because it is the air movements that we observe from the ground which constitute essential components of our weather. But, as already mentioned, if we wish to compute air motion relative to the rotating system, accelerations and associated inertia forces associated with the rotation must be included in the equations of motion. [It must be emphasised that in many common fluid dynamic phenomena observed on Earth, the accelerations of fluid particles relative to the local surface of the Earth greatly exceed that due to Earth rotation, so that the latter can be ignored. But this is not true in the cases of large-scale, slow-moving atmospheric and oceanic flows.] For the purposes of analysing atmospheric dynamics, it turns out that a frame of reference having its origin at the centre of the Earth, and which does not rotate with the Earth, is sufficiently 'inertial'. Of course, it is not strictly true that the centre of the Earth does not itself accelerate because it traces out an elliptical orbit around the Sun. But, for the purpose of analysing and explaining the movements of air in the atmosphere, this contribution to acceleration has negligible effect.

Consider a tree standing beside a merry-go-round (for the sake of simplicity let us temporarily assume that the ground in which the tree stands constitutes an inertial reference system). To an observer standing on the edge of a rotating merry-go-round, the tree (together with the rest of everything in view except for the merry-go-round platform and handrails) is moving. The tree possesses velocity and, crucially, the direction of that velocity vector is seen as changing with time. As we know from the example of the tyre in Chapter 3 (see Fig. 3.1), variation in time of a velocity vector constitutes acceleration: hence, the tree (and its surroundings) is seen to be accelerating. Therefore, if the observer wishes to apply Newton's Laws of Motion to what he or she observes, he or she must assume that the tree is subject to a force to make it accelerate. This is termed 'Coriolis acceleration': when multiplied by the mass (or inertia) of an accelerating element, it is termed 'Coriolis force'. It is clear that this 'force' is an inertia force that originates simply because of the choice of a non-inertial frame of reference that is rotating and translating: *it is not a physical force that can do work and change kinetic energy.* These effects will be explicitly discussed later to explain circular forms of atmospheric air motion such as cyclones. A warning: the concepts of Coriolis acceleration and Coriolis force are difficult to comprehend, as demonstrated by the existence of a plethora of different explanations and demos.

The Coriolis effect in two-dimensional (plane) motion is sometimes illustrated in terms of the model of someone attempting to roll a ball outwards along a radius of a rotating circular platform such as a merry-go-round. It is explained that the ball will take a curved path relative to the rotating system, but will be seen by an onlooker sitting on the branch of a tree looking down from above (in his or her inertial frame) to move outwards radially along a straight line, as illustrated by Fig. 5.1.

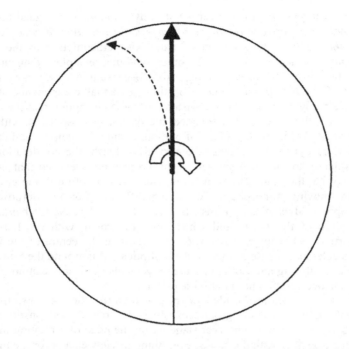

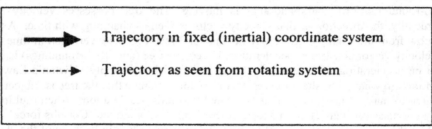

➡	Trajectory in fixed (inertial) coordinate system
┄┄┄➤	Trajectory as seen from rotating system

Fig. 5.1 Simple Coriolis effect

This model is dynamically flawed. If one places a smooth ball, such as a pool ball, on a tablecloth and fairly quickly, but steadily, pulls the cloth off the table, the ball will initially slip and undergo both translational and rotational acceleration due to the friction force between the ball and the cloth. But, once the rotational speed of the ball is such that its peripheral speed equals that of the tablecloth, it will roll and remain in a (more or less) fixed position relative to the table [Try it]. If a ball were to roll (without slipping) outwards in a radial direction on a *rotating* platform, its transverse rotational speed would have to continuously increase to match the local speed of the platform, i.e. to undergo transverse rotational acceleration. This requires the action of a friction between it and the platform. This frictional force will impart tangential acceleration to the ball, which will not therefore trace out a straight line as seen from the non-rotating (inertial) frame of reference represented by the onlooker overhead. If we substitute the ball by an ice hockey puck and cover the rotating platform with an almost frictionless Teflon (PTFE) surface, the scenario of Fig. 5.1 will be more nearly realised. To someone rotating with the merry-go-round, it would appear that the puck is subject to a force acting in a direction perpendicular to the velocity vector of the puck.

In dealing with atmospheric airflows, we who are 'attached' to the Earth's surface, are concerned principally with air movements relative to that surface. The fact that the Earth is (almost) spherical makes the Coriolis effect three-dimensional. For example, at any latitude except the equator, the centripetal acceleration vector of an element of air towards the Earth's axis has components in both the direction perpendicular to the local surface of the Earth (downwards) and parallel to the surface (along the floor) as illustrated by Fig. 5.2. The latter component is the more important in atmospheric dynamics. This feature, together with the fact that air is subject to a range of different physical influences, renders the effects much more complex and difficult to visualise than those associated with the merry-go-round. The concepts of Coriolis acceleration and the associated Coriolis force are universally and validly used by meteorological scientists and mathematicians to allow the motion of the air (and oceans) relative to the Earth's surface to be mathematically expressed, analysed, graphically represented and understood in terms of classical Newtonian mechanics. Since Coriolis did not consider applying his analysis to the motions of the atmosphere or the oceans, it could be argued that the phenomenon should, in justice, be renamed the 'Ferrel/Coriolis' effect.

5.4 MOTION OF ATMOSPHERIC AIR

The motion of atmospheric air is subject to four main influences: gravitational forces, pressure gradients, temperature gradients and viscous forces. As indicated in the previous section, because we observe the motion from a moving platform on the surface of the rotating Earth, the Coriolis effect must also be taken into account, as further discussed in relation to cyclonic flows. Centrifugal effects, which are proportional to the square of the Earth's rotational speed, are too small to be of consequence compared with the influence of gravity. Meteorologists and atmospheric scientists identify a number of generic forms of air motion that are distinguished by their spatial scales, typical speeds, directional characteristics, lifetimes and driving influences. The motions of atmospheric air are extremely complex, so the following sections present greatly simplified qualitative accounts of the principal phenomena.

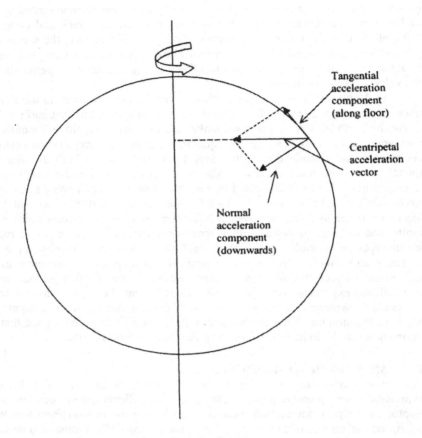

Fig. 5.2 Acceleration in rotating systems

5.5   CYCLONES AND ANTI-CYCLONES IN THE TEMPERATE REGIONS

The forms of meteorological phenomena with which readers in temperate regions will be most familiar are the cyclone (low pressure) and anti-cyclone (high pressure) systems that are generally associated with cloudy and clear skies, respectively. Temperate cyclones are large-scale vortex-like movements of air around a central region of low pressure (depressions) in which the Coriolis effect is dominant. They are largely absent from regions within about 30° N/S latitude of the Equator where the Coriolis effect is weak. They typically extend over hundreds of miles and last for a few days. Plots of cyclones as seen on television weather forecasts, exemplified by Fig. 5.3, show approximately circulars pattern of isobars (contours of equal pressure). Each isobar represents a different pressure, with the lowest pressure at the centre. The local pressure *gradients,* which determine the forces on fluid elements from the surrounding fluid, are therefore *perpendicular* to the local isobars; the closer the isobars, the greater the pressure gradient. However the cyclonic wind flows in directions almost *tangential* (parallel) to the isobars, rather than in directions perpendicular to them that the chapter on aerodynamics would suggest. The closer the spacing of the isobars, the stronger the wind. Cyclonic winds can be very strong and can cause considerable damage to vegetation. Anti-cyclones are similar to cyclones but the pressure increases towards the centre, and they are associated with clear, sunny weather. They are more limited in strength and commonly do not exhibit such a clear circulatory pattern of wind as cyclones.

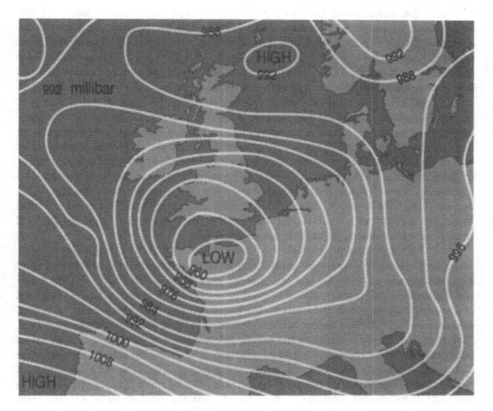

Fig. 5.3 Isobars around a depression

The apparent ability of the air involved in the cyclonic process to defy Newton's Second Law of Motion by moving parallel to the isobars is most concisely explained in terms of fairly advanced mathematical representations of the effects of Earth rotation on the motions of air relative to the Earth's surface. [See the article by J. Price in 'Further reading'.] It is very challenging to attempt to explain the Coriolis effect on the airflow in a temperate cyclone in words: however, here goes. A cyclone is associated with the establishment of a localised region of low pressure, usually caused by spatial variations of temperature in the surrounding atmosphere. The Coriolis effect is initiated by the 'horizontal' acceleration of air from all azimuthal directions towards the region of low pressure, driven by the radially directed pressure gradient distribution. These accelerations are related by Newton's Second Law of Motion to the local pressure gradients that lie at right angles to the local isobars. The whole cyclonic system rotates with the Earth and consequently, if fluid continued to flow in a radial direction, the resulting fluid velocity vector would rotate relative to an inertial frame of reference. However, changes of direction of a velocity vector constitutes acceleration in a direction perpendicular to the vector, as we have seen in Chapter 1. But there exist no forces on the fluid particles to generate accelerations in these directions. Consequently, the inertia of the fluid particles cause their velocities to follow a curved path *relative to the Earth's surface* such that it 'cancels' the acceleration that would otherwise imposed by the rotation of the atmosphere. In whatever direction (East, South, North or West) a fluid particle is driven by the pressure gradient, this inertial 'resistance' causes the velocity direction to continuously curve to the right of its current direction This surprising fact is a result of the sphericity of the Earth by which the 'plane' of a cyclonic flow regime lies at an angle to the axis of rotation of the Earth. This accounts for the weakness of the Coriolis effect at latitudes close to the Equator. On the basis of the assumption by the Earth-bound observer that Newton's Second law applies to the fluid motions as observed by him or her, it is necessary to assume that a force is acting to account for the curved trajectory. This is the Coriolis force and the acceleration component that it 'causes' is the Coriolis acceleration. Its magnitude is proportional to product of the speed of a fluid particle *relative to a rotating frame* and the rate of rotation of the frame. The Coriolis force is not a physical force in the conventional sense because it cannot do work to change the kinetic energy of a particle. It is an inertia force; an artefact which allows Newton's Law to be applied to motions defined relative to a non-inertial, rotating frame of reference.

As the process progresses, the turning of the particle velocity is sufficient to bring its acceleration vector to oppose the radial acceleration imposed by pressure gradient associated with the isobars, thereby cancelling it. The air is free to move parallel to the local isobars, and continues to circulate for periods of days, although it does drift slowly across the isobars. The process is illustrated by the very idealised diagram in Fig. 5.4. The centre circle represents the flow in the established cyclone. The arrows pointing inwards to the centre circle represent the component of radial acceleration due to the radial pressure gradient. The opposing arrows represent the balancing Coriolis component of radial acceleration. The Coriolis component of acceleration is represented by the short arrows. The combination turns the flow to the right of its current direction, and ultimately into the central circle. As seen from above, cyclonic winds in the northern hemisphere rotate in an anti-clockwise direction, as illustrated by the example of the flow around a low pressure area over Iceland seen in Fig. 5.5; and in a clockwise direction in the southern hemisphere. The technical term for this form of flow is 'geostrophic' which is derived from the Greek word 'strophē' for 'turning'.

5.6   HADLEY CIRCULATION

Solar heating of the atmosphere is strongest in the equatorial belt. The heated air rises and is displaced by cooler and denser air from regions to the north and south. At heights of between 10 and 15 kilometres, the displaced heated air flows towards the poles. As it cools, it falls and returns at low level to complete the journey, as shown by Fig. 5.6, in which the darker areas near the Equator and the west coast of South America that have negative numerical labels indicate rising air and the lighter areas having positive numerical labels indicate falling air. This is called an 'overturning' circulation and the system is called a 'Hadley cell'. As the displaced air reaches the tropopause, the Coriolis effect turns the winds eastwards in the both hemispheres, creating high speed jet streams that flow from west to east (see below). This phenomenon can also be understood in terms of the angular momentum of the rising flow. In simple terms, the angular momentum of a body that traces out a circular path around an axis of rotation is proportional product of its speed and the distance to the axis of rotation. If no turning torque acts on it, its angular momentum is conserved (constant). The spinning skater presents a common

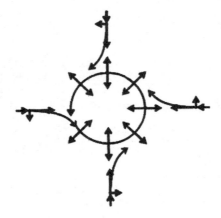

Fig. 5.4 Diagrammatic representation of the Coriolis effect around a depression

Fig. 5.5 Flow around a low pressure region

example. She speeds up as she moves her arms towards her axis of rotation. You can demonstrate it for yourself by asking a friend to rotate you in a computer chair and then thrusting out, or drawing in, your legs. As the air moves towards a pole, it gets closer to the Earth's axis of rotation and therefore, because that angular momentum is conserved, it speeds up. The returning flow is redirected to the west and slows down for the same reason: it constitutes the 'trade wind'.

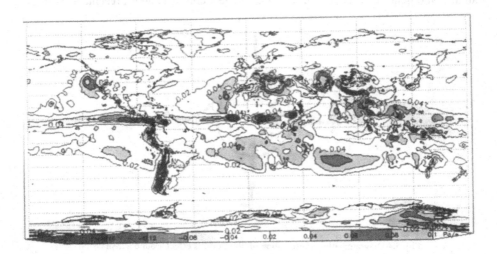

Fig. 5.6 Hadley circulation

5.7   JET STREAMS

In November 1944, a fleet of American B29 bombers headed east to bomb the north west region of Tokyo in Japan. They had on board the latest in bomb sight systems that compensated for the speed of the aircraft over the ground and allowed them to aim and drop bombs from a height of over thirty thousand feet, which was well above the height at which Japanese fighters could operate for any useful length of time. The nominal cruising speed of the planes at that height was about 340 mph. On reaching the target area they climbed to thirty seven thousand feet, turned and headed west on a bombing run. But they could not lock on to the targets because they repeatedly overshot them. The radar operator checked the ground speed and estimated that the planes were subject to a 125 mph tailwind; the computed ground speed was about 480 mph. They dropped their bomb loads randomly and returned to base where the crews reported the problem in concert. The General in charge of the operation did not believe them. However, he eventually had to accept their accounts because they were confirmed by an operations officer who was a passenger on board one of the aircraft; and the US military weather forecasters had computed the wind speed to be about 168 knots at the operational height and target location. The aircrews had encountered what they termed 'a river of air'. High altitude bombing missions in that region were abandoned.

What they had encountered became known as the 'jet stream'. What they did not know was that a Japanese meteorologist, Wasaburo Ooishi, had discovered and measured the speed of jet stream over Japan twenty years earlier using hydrogen-filled balloons observed from the ground through telescopes. In 1926, he had tried to communicate his discovery to the world's meteorologists by writing a paper in Esperanto: but it had been ignored. During WW2, the Japanese exploited the jet stream to send about one thousand balloons carrying explosives across the Pacific Ocean to the west coast of the United States. They were largely ineffective, but one sadly killed a group of children on a Sunday school picnic trip in Oregon in 1945. The American meteorologists now knew that the jet stream operated over distances in excess of 6000 miles.

There are two major jet streams that flow in an easterly direction around the Earth where the troposphere meets the stratosphere at an altitude of about ten kilometres, between latitudes of about 30° and 70°, one in each of the hemispheres. As shown in Chapter 1, the temperature falls with height in the former atmospheric layer and rises with height in the latter. These are known as the 'polar jet streams' because they are created by the interaction between warm air rising from the tropics and cooler air moving from the polar regions. The associated pressure gradients, together with the effect of the rotation of the Earth, combine to produce streams of air that can reach speeds as high as 200 mph, although they generally flow at speeds between 50 and 80 mph. They are typically one to two hundred kilometres in width and about one to two kilometres in depth. The Northern polar jet stream originates over North Africa and flows across the Middle East; then over mid-Asia, causing the familiar plume of snow blowing off Mount Everest; then out over the Pacific to North America and finally over the North Atlantic Ocean, making landfall in Northern Europe, frequently over the United Kingdom, where it dies out. The Southern jet stream generally flows over the southern areas of South America, South Africa and Australia. There are other, generally weaker, jet streams that flow in equatorial latitudes. Here we shall concentrate on the behaviour and effects of the northern polar jet stream.

The paths of jet streams are highly irregular, as illustrated by Fig. 5.7. They have profound effects on the world's weather system. One of the most important features of the northern jet stream is the capacity to act as a channel for cyclonic systems. For example, it steers cyclonic depressions and associated storms across the north Atlantic and brings successive spells of mild, wet weather to north west Europe. Its propensity to snake, together with the then less advanced forecasting technology, led to a failure to forecast the path of the jet stream in October 1987 which left residents of the south coast of England and northern France unaware that it had turned northwards from its original trajectory towards Spain. The ensuing storm destroyed 15 million trees in the south of England and caused widespread damage to property.

Strangely, an absence of the jet stream can also cause extreme climatic conditions. The jet stream that terminates in the region of the British Isles can be prevented from making landfall by a phenomenon known as a 'blocking event', caused by strong cold winds arriving from the east. This was responsible for the periods of very cold, snowy winters experienced by Northern Europe during the 1950s and early 1960s. In the late 1990s its path over the United States shifted to the north. Its absence was responsible for allowing high pressure systems to dominate

Fig. 5.7 Jet streams

over a period of years causing prolonged drought in some north western states, conditions favourable to the ignition and rapid spread of a succession of forest fires which endangered life, damaged property and ravaged woodland, and which were economically damaging. The very wet summers and widespread flooding experienced by the British Isles in the 'summers' of 2007 and 2008 are attributed to a southward shift of the jet stream in this region. However, there is some evidence that, in the medium term, a northward shift will occur over Europe, quite possibly caused by global warming. This will probably have severe repercussions for the future climate of northern Europe - Britain in particular. The most likely scenarios are warm, wet, stormy winters and very hot summers accompanied by brief periods of very heavy rain. The jet stream is also involved in the generation of tornadoes over the central plains of United States which annually cause fatalities and millions of dollars' worth of damage to property, as in Oklahoma City in 1999.

On the positive side, the northern jet stream is of considerable economic value to airline companies that operate routes over the North Atlantic. By flying in the jet stream, pilots can reduce the eastward bound journey from the eastern US to Europe time by more than one hour compared with the unaided trip. However, on the return trip, they must avoid the jet stream, which requires them to take the 'Great Circle' route up to Iceland and then south west to the US.

5.8   OTHER VORTICAL FLOWS

Atmospheric phenomena include a number of different types of vortical systems in addition to temperate geostrophic flows. These include tropical cyclones, which are called 'hurricanes' in the western Atlantic and 'typhoons' in west Asia, tornadoes, dust devils and land-sea winds. The two major kinematic features of the flows are the Coriolis and centripetal accelerations, which are conventionally represented by Coriolis and centrifugal (or inertial) forces. These types of flow are characterised by the ratio of these two influences in terms of a non-dimensional parameter termed the Rossby number, denoted by R_0. If R_0 is much less than unity, Coriolis forces dominate; otherwise, centrifugal forces dominate. For tropical cyclones, R_0 is about 0.1. For land-sea winds, R_0 is about unity. For tornadoes and dust devils, R_0 is of the order of 1000. The vortex associated with the drain plug of a bath also has a value of R_0 of about 1000. This means that the direction of rotation has nothing to do with Coriolis or the particular hemisphere: it is determined by the residual movement of the water before the plug is pulled. So-called 'demonstrations' of the influence of the position of the plug in relation to the equator are fraudulent.

Fig. 5.8 Hurricane Katrina

Tropical cyclones form above warm tropical oceans where the humidity is very high. Strong upward thermal convection creates storm clouds up to 25 km high in which the moist air is transported into regions of lower temperature where the vapour condenses releasing latent heat energy. The Coriolis effect forms a cyclone which strengthens in passage over a warm ocean to generate a vortex storm, with air speeds up to 80 m/s, as evidenced by hurricane Katrina that devastated New Orleans in 2006 (Fig. 5.8). Tropical cyclones are typically 500 km in diameter and last for up to 10 days. The air pressure at the centre is only about nine tenths of the average sea level pressure which allows the rotational flow, undergoing very strong radial acceleration, to exist, as explained in Chapter 3. Tornadoes are small-scale cyclones that take too many different forms to be described in detail here. They are typically produced by local thermal convection in the form of a concentrated tube (see Fig. 5.9) that stretches from the ground and terminates in a cumulus or cumulonimbus cloud (see the section on clouds below). They are typically one kilometre in diameter, lasting about one hour and generating wind speeds of up to 100 m/s. In both cases, the strong central upward flow induces compensating, inwardly radial, largely horizontal flow near the ground. This is the wind that does the damage to terrestrial structures. Over the ocean, tornadoes cause rising columns of water called waterspouts. Dust devils are miniature tornadoes that do little harm except for redistributing ploughed farmland.

Fig. 5.9 A tornado

Land-sea winds are vertically circulating flows usually occurring on coasts. During a sunny day, the air lying over the land heats up more than that over the water, particularly if it slopes to face the midday sun. This warm air rises, being displaced by cooler air from the ocean to form an on-shore wind, and falling over the sea as it meets cooler air above. At night, the land cools faster than the air (as with ground frost), and the circulation direction is reversed.

5.9   THE IMPORTANCE AND EFFECTS OF WIND

Winds perform many functions that are of vital importance to human existence. They transport excessive heat from the equatorial regions towards the temperate regions. By means of viscous frictional forces and pressure forces on the resulting wave peaks, winds drive ocean currents. Wind-driven waves are agents in the interchange of atmospheric gases with the oceans which plays a very important role in controlling the balance of oxygen and carbon dioxide in the atmosphere. Winds drive rain clouds across the land. They disperse pollutants, seeds, bacteria, fungi and algae. They are beginning to provide a significant contribution to the generation of electric power both by driving wind turbines and through the generation and propulsion of water waves. They dry the washing. They can also threaten human life and do extensive damage to natural and man-made structures, both directly and by fanning forest fires. A little known effect of wind is to vary the rate of rotation of the Earth, so that one day can vary by as much as a few milliseconds. The cause is the force applied by the wind to the major mountains of the Earth, the direction of which varies from day to day.

The meaning and nature of laminar and turbulent boundary layers have been addressed in Chapter 3. When wind blows over the surface of the Earth, a boundary layer is generated. However, it is substantially different from that generated on a smooth surface, such as an aircraft wing or the inside surface of a pipe, because of the many forms of irregularity presented by topographical features such as hills, trees and hedges, buildings, and the multitude of other obstacles to free airflow. The boundary layer is turbulent and strong mixing occurs. This is beneficial in dispersing gaseous pollutants and promoting the exchange of oxygen and carbon dioxide at the surface of the sea, but undesirable in causing damage to vegetation and man-made structures and in causing soil erosion. The average vertical extent of the boundary layer is 457 m for large cities, 366 m for suburbs, 274 m for open terrain and 213 m for open sea. The boundary layer flow is three-dimensional and there can be significant differences in flow direction between it and the 'free' flow above it, known as the 'Ekman spiral effect', which partly accounts for the fact that the directions of wind felt at ground level can differ from that observed from the motion of clouds. As explained in Chapter 4, the wind's boundary layer has a significant effect on the direction and unsteadiness of propagation of sound close to the ground through the phenomenon of refraction .

5.11   CLOUDS AND THUNDER STORMS

As you know, there are many different forms of cloud. The conditions and mechanisms of generation, evolution and decay are too varied and complex to be covered in detail in this small book. A typical small cumulus cloud has a billowy, cauliflower-like, appearance having a volume of about one cubic kilometre and a mass of one thousand million kilograms. Since clouds are formed of water vapour, dust particles and, sometimes, ice crystals, they are denser than the surrounding air; but they appear to float! How come? In fact, they don't 'float' due to buoyancy like a boat or a helium balloon.

The most important cause of cloud formation is the presence of rising air. As mentioned in Chapter 3 in relation to gliders and large birds, these regions are known as thermals. Interestingly, balloonists and paraglider pilots use sound from sources on the ground as one means of detecting the presence of thermals. Because the warm air is rising and the speed of sound is higher than in the surrounding cooler air, sound is refracted into a thermal and channelled upwards. The volume of water that can be held per cubic metre of air in the form of invisible water vapour (a gas) increases with temperature. If warm air is caused to cool, the molecules of water vapour condense (aggregate) into fine droplets of liquid that are visible, as exemplified by exhaled breath on a very cold day and suspensions in the air as mist and fog. Warm air may be caused to rise and cool naturally by three basic mechanisms. Air near the surface of the Earth collects heat from ground that is heated by the sun. Over regions of ground surrounded by cooler regions the more buoyant air rises to be replaced by cooler air. Wind may be forced upwards onto cooler regions by topographical features such as hills and mountains. A warm wind may encounter a region of cooler air and rise above it as in a weather front.

Water droplets in a cloud are typically of the order of one to ten microns in diameter. If cooled sufficiently they will aggregate into raindrops. There are between ten and one hundred droplets in the average raindrop. Gravitational force causes the droplets to fall *relative* to the rising air; the relative speed of fall is limited because the aerodynamic drag imposed by the viscosity of the passing air is proportional to the relative velocity (very low Reynolds number) and they reach a terminal relative speed. If the updraft speed is greater than this, the particles move upwards relative to the ground. However, the weight of a raindrop is proportional to the cube of its diameter whereas viscous drag is proportional to the square of its diameter. So, if sufficiently large drops form, they will fall as rain.

As any air traveller will know, clouds have upper and (usually) lower height boundaries. The upper one is normally ragged and the shape of the top of a cumulus cloud can be observed to change continuously in a kind of 'boiling' motion as water droplets reach the top. The lower boundary is usually more clearly defined and steady. Warm air cannot rise for ever and, as it rises, it must be replaced by colder air. As the rising warm currents of air reach regions of the surrounding atmosphere that have much lower temperature, they lose heat and ultimately start to return to earth under the influence of gravity, taking water droplets along for the ride. Observation of a group of clouds of similar form in fairly quiet weather conditions will reveal that their lower boundaries all lie at roughly the same height above ground, called the 'cloud base'. It's always comforting to drop through the cloud base and see the ground on which your aircraft is going to land. This is the height at which the temperature of the surrounding atmosphere is high enough to turn the droplets to invisible vapour.

Some forms of cumulus clouds reach very great heights of up to about 17,000 metres; they are technically called 'cumulonimbus' clouds and, in the vernacular, 'thunder' clouds. They form on hot and humid days when the atmospheric temperature falls rapidly at high altitude. They contain vast amounts of moisture, together with ice crystals and hailstones, and strong up and down drafts of air. There appear to be a number of different theories about the process by which positive and negative electrical charges are generated within a thundercloud, perhaps by collision between upwardly and downwardly moving ice crystals and hailstones. However, it is known that the ice crystals acquire positive charges and are elevated by updrafts, whereas the hailstones acquire negative charges and fall in downdrafts to congregate in the lower region of the cloud. A positive charge builds up on the ground beneath the cloud.

Once the negative charge becomes large enough to overcome the air resistance, a flow of negative charge, called a 'stepped leader', rushes towards the ground and a flow of positive charge leaves the ground to meet it. When the two leaders meet, the main discharge, called the 'return stroke' travels from *ground to cloud* (Fig. 5.10). This is the event that causes most of the damage caused by lightning. The temperature of this discharge is about 30,000°C. The surrounding air is suddenly heated, ionised and expands extremely rapidly. This constitutes an extended Category 1 acoustic source. The radiated disturbance takes the form of a shock wave which, unlike a normal sound wave, changes its form as it propagates, so that it is heard as a sharp crack by nearby listeners, but as more of a boom by those further away. The repetitions that are often heard are echoes of the sound from surrounding hills. Lightning can also travel within a thunder cloud and from cloud to cloud. Surprisingly, it transpires that nobody knows for certain how lightning conductors work: fortunately, most of them do - an excellent example of a triumph of empiricism over theory.

Fig. 5.10 Lightning strikes

6

Air Technology: Uses and Applications

When the people heard the sound of the trumpet, and the people shouted with a great shout, that the wall fell down flat......

The Bible Joshua ch. 10, v. 12

6.1   INTRODUCTION

The properties and behaviour of air are exploited in a myriad of ways. This chapter presents brief descriptions of a small selection of the less well-known uses and applications of air, and is in no way exhaustive. However, it opens with a section on the well-known subject of musical wind instruments because, although many readers will have personal experience of these instruments, many will not be familiar with the physical processes by which they produce sound.

6.2   MUSICAL WIND INSTRUMENTS

Musical instruments are traditionally categorised as follows. Aerophones: these produce sound by modulating airflow. Idiophones: these are made of naturally sonorous solid materials that are excited in a number of different ways, particularly by striking. Membranophones: these produce sound by the vibration of stretched membranes or skins. Cordophones: these comprise stretched strings which are vibrated, but which radiate little sound themselves. The vibration is transmitted to sound 'box' to which they are attached and which radiates almost all of the sound. The aerophones group includes flutes, panpipes, recorders, clarinets, saxophones, oboes, bassoons, bagpipes, trumpets, trombones, horns, tubas, cornets, harmonicas, concertinas, accordions, organs and bull-roarers. This section briefly describes the generic sound generation mechanisms of common tubular Western orchestral wind instruments that form a subset of the aerophones group.

Before dealing with the driving airflow, we digress to explain the phenomenon of acoustic resonance that is common to almost all aerophones. The simplest form of acoustic resonator is the Helmholtz resonator described in Chapter 4. A one-litre plastic drink bottle that has *distinct neck* forms a good Helmholtz resonator. If you blow *gently across* the mouth of the neck of the bottle, a tone will be heard. The bottle will be felt to vibrate as the pressure in the contained air oscillates. Those you who have a piano might like to try to match the tone on the keyboard. Now, if you place your mouth above and close to, but not closing, the neck, and sweep a sung tone ('aaah') across a small range encompassing the blown tone, the bottle will be felt to vibrate strongly when you hit *resonance*, in which the excitation frequency equals the natural frequency. These two cases of excitation are quite different in nature. The vocal excitation is independent of the acoustic response of the resonator. But the aerodynamic excitation of the breath is strongly linked to the response because the driving agent - the airflow - is fluid-dynamically *coupled* to the oscillatory motion of the air in the neck. As it moves across the bottle's mouth, the turbulent jet issuing from your mouth forms a shear layer (see Chapter 3) which is unstable and very susceptible to disturbance. This unsteady flow disturbs the mass of air in the neck which moves in and out and reacts with the air spring in the body of the bottle; causing sustained resonance to begin. The oscillating air in the neck disturbs the air jet, and *vice versa*. This is a simple example of 'aeroacoustic' coupling, a phenomenon that is central to the mechanism of sound generation by most wind instruments.

The majority of Occidental orchestral wind instruments are basically tubular in form, although the bore is not usually uniform, and the tubes of many wind instruments terminate in an expanded horn. In order to explain the acoustic behaviour of tubes, consider a uniform tube that is closed at one end and open at the other. You may observe the acoustic response behaviour in a similar manner to that employed for the Helmholtz resonator. Take a flexible vacuum cleaner pipe, or other very flexible plastic tube of at least 1 m in length. Place one end firmly over an ear. With you mouth close to, but not quite closing, the open end, sing *gently* (to avoid hearing damage) and sweep the pitch of your voice *slowly* upwards. You will hear the sound level in your ear rise to maxima at a number of different pitches, and may also feel associated peaks in tube vibration. If you have a musical instrument you will be able to match the those frequencies to the pitch of your voice and observe that they are (very closely) odd multiples of the fundamental. These are the resonance frequencies of the tube which correspond to the acoustic natural frequencies. They are known variously as 'overtones', 'harmonics' and 'partials'.

So, how are these resonances produced? Unlike the air in a Helmholtz resonator, the air in a tube does not behave as a simple mass-spring system at its resonance frequencies. Imagine that the open end of the tube is suddenly exposed to single frequency sound from a nearby loudspeaker. A sound wave of that frequency will travel along the tube until it hits the closed end from where it will reflect (reverse the direction of propagation) and travel back towards the open end. Surprisingly, sound waves having wavelengths that are much longer than the circumference of a tube in which they are travelling will be largely, but not completely, *reflected* at the *open* end. This is because the associated movement of air in and out of the opening acts rather like the cone of a loudspeaker, and we know that small loudspeakers do not radiate bass (long wavelength) sound effectively. Consequently, having reflected from the open end, these slightly attenuated waves repeat their forward journey down the tube, where they reflect again from the closed end, and so on. At certain excitation frequencies, all the waves in the tube reflected from the open end will closely match those sent directly down the tube by the loudspeaker (technically, this match is termed 'phase coincidence') and resonance occurs. As a consequence, the response builds up and grows to maximum that is limited only by losses of energy to viscous friction in the acoustic boundary layer the walls and by radiation of sound energy from the open end: these mechanisms are weak, but still limiting.

As the frequency increases, an increasing proportion of the energy of sound waves reflected from the closed end and travelling towards the open end is radiated away into the external air at the opening, which weakens the subsequent reflection back down the tube. This radiation constitutes a form of damping (energy loss mechanism). Consequently, the strength of the resonant responses of the harmonics decreases with increase in the harmonic number until ultimately no higher harmonics, and no corresponding tones, are generated. Paradoxically, it is necessary to 'bottle up' most of the reflected sound energy inside a wind instrument to produce the relatively weak leakage of sound that we hear. For example, if one were to attach a good acoustic horn to the top of a pipe of a flue organ, which would allow much of the energy of approaching sound waves to escape into the surrounding air, the pipe would not sound, because the driving mechanism depends upon a strong feedback of sound from the open end to maintain and strengthen the oscillatory flow in and out of the flue at the bottom.

Associated with each harmonic resonance is a 'standing wave' in the tube. This is a spatially sinusoidal distribution of sound pressure and particle velocity formed by the superposition (interference) of the waves travelling in opposite directions. At the lowest (fundamental) resonance frequency, the acoustic wavelength is very close to four times the length of the tube; the sound pressure varies smoothly from a maximum at a closed end to close to zero at the open end; the acoustic particle velocity is zero at the closed end and close to maximum at an open end. The wavelength at the next highest harmonic cannot be twice the length of the tube because this would require the acoustic pressure to be maximum at both ends - but it must be almost zero at the open end. So the next harmonic frequency is three times that of the fundamental, and the acoustic wavelength is very close to four thirds of the length of the tube. The same conditions as above obtain at the ends, but there is a pressure minimum (nearly zero) at a position two-thirds of the length from the open end.. The higher harmonics, that occur only at *odd multiples* of the fundamental frequency of such a tube, have an increasing number of these minima spaced along the length of the tube, as shown in Fig. 6.1 which shows the standing wave pattern of sound pressure with the open end on the left. To be more precise, the reflection from an open end does not occur precisely at the geometric end. A small amount of air just outside the open end oscillates with the wave, and reflection effectively takes place at small distance beyond the end that is weakly dependent upon frequency. This is the main reason why the harmonics of simple, uniform, open-ended tubes are not exactly evenly spaced in frequency.

Musical 'pitch' is the subjective perception and attribution of frequency. The pitch of a steadily played note is usually determined by the frequency of its fundamental harmonic, even if there is far less sound energy in this than in higher harmonics. The classical example is that of the oboe of which the energies of the fundamental harmonics of the notes in its lower register are up to 300 times smaller

Fig. 6.1 Sound pressure distributions of the first four harmonics of a tube open at the left hand end and closed at the right hand end

than those of the strongest higher harmonics: in principle, they cannot actually be heard! The brain appears to infer the pitch by extrapolating downwards from the pattern formed by the regular sequence in frequency of the higher harmonics. For example the inferred fundamental of the sequence 690, 920, 1150, 1380 Hz is 230 Hz.

Wind instruments behave acoustically in a much more complicated fashion than the simple tube previously discussed, partly because their bores are mostly not uniform and many have side holes, and partly because the excitation mechanisms involve interaction between the internal sound field and the aerodynamic driving mechanism. Consequently, both even and odd harmonics can be excited; but the relative strengths of the harmonics, which largely determines the *timbre*, or acoustic character, of wind instruments, change with the type of instrument and the register in which it is played; the register is a part of its frequency range that has a distinctive tonal quality. For example, clarinet sound is characterised by the weakness of the even harmonics in the lower registers, as shown by the frequency spectra in Fig. 6.2.

The process of initial build up to a steady note (known as the 'attack') is determined by the relative strengths and rates of build up of the note's harmonics which depend upon instrument form and quality, and very much on performer skill. Attack is extremely important to the auditory process of identification of instrument type. If the attack is electronically removed from recordings of individual notes played on various instruments it is very difficult to distinguish certain pairs of instruments of quite different form and mechanism.

The resonances of wind instruments are excited by oscillatory modulation of an input airflow. This is effected in a number of different ways. Many wind instruments incorporate mechanical valves called 'reeds'. In clarinets (Fig. 6.3(a)), oboes (Fig. 6.3(b)) and reed organ pipes (Fig. 6.3(c)), blowing pressure tends to close the valve and the internal sound field opens them. Harmonium reeds are like vocal cords in opening in response to blowing pressure. The valves of unreeded brass instruments such as the trumpet are formed by the player's lips as shown in Fig. 6.3(d). In other reedless instruments such as the side-blown flute (Fig. 6.4(b)), recorder (Fig. 6.4(a)) and flue organs (Fig. 6.5), a jet of air is blown against the sharp edge of a thin plate or lip of an opening. Figs. 6.3 and 6.4 illustrate the two generic forms. Reeds have their own natural frequencies, but in all cases of tubular instruments, dynamic interaction between the resonant sound field within instrument

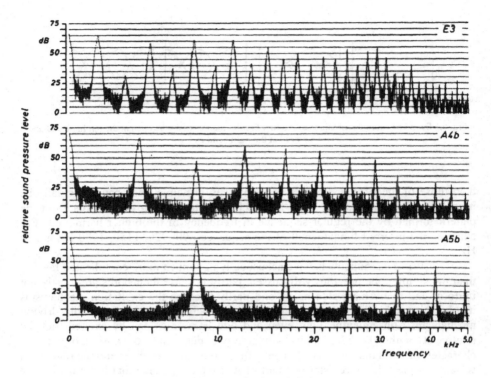

Fig. 6.2 Spectra of various clarinet notes

and the driving mechanism is fundamental to the process of generation of musical notes. This is not the case for tubeless instruments that incorporate free reeds such as the accordion and concertina. Most reeded and brass instruments are effectively acoustically closed at the driving end. The opening and closing of reeds and lips is influenced primarily by the local sound pressure together with the Bernoulli effect (see Chapter 3) of the flow through the restriction formed by the valves.

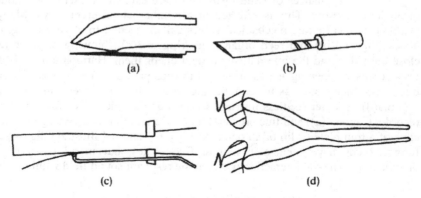

Fig. 6.3 A selection of valved wind instruments

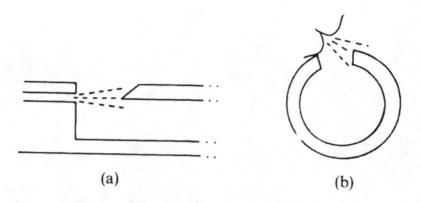

(a)   (b)

Fig. 6.4 A selection of unvalved wind instruments

A jet of air emerging from a thin slit is fundamentally unstable and if directed at the sharp edge of a plate or sharp lip is easily 'persuaded' by aeroacoustic interaction with sound in a pipe to deviate in an oscillatory fashion so as to flow principally over one side or the other, as illustrated by the sequence of photos in Fig. 6.5. (also see the section on Fluidics below). In the cases of side-blown flutes, pipe organs and similarly driven instruments, the driving end is acoustically open, and it is primarily the oscillatory acoustic particle velocity (actually volume velocity) in and out of the opening that couples with the oscillatory jet flow. Sound waves travelling in a tube are not only reflected at an open end, but also at the location of side holes in the tube, from which sound also radiates. By opening and closing side holes, wind players of clarinets, oboes, recorders and other such keyed or fingered instruments can select the pitch of the note played. The bores of most wind instruments are not uniform, partly to produce more evenly responding harmonics, but also to stamp the makers individual timbre on the instrument. Brass instruments radiate only from their open ends. Termination by an expansion section (horn) allows designers to manipulate the frequencies and relative strengths of the harmonics and the directional radiation characteristics.

The aeroacoustic interactions described above are essentially nonlinear. This means that the relative responses of the various harmonics that determine the 'timbre' depend upon the strength of the blown airflow. Instruments can be 'overblown' to increase their pitch range. The sound of some wind instruments, particularly flutes, is not purely harmonic because noise associated with the blowing flow is also distinctly heard and is an intrinsic feature of the flute sound. Sound waves travelling in a tube are not only reflected at an open end, but also at the location of side holes in the tube, from which sound also radiates. By opening and closing side holes, wind players of clarinets, oboes, recorders and other such keyed or fingered instruments can select the pitch of the note played. The bores of most wind instruments are not uniform, partly to produce more evenly responding harmonics, but also to stamp the makers individual timbre on the instrument. Brass

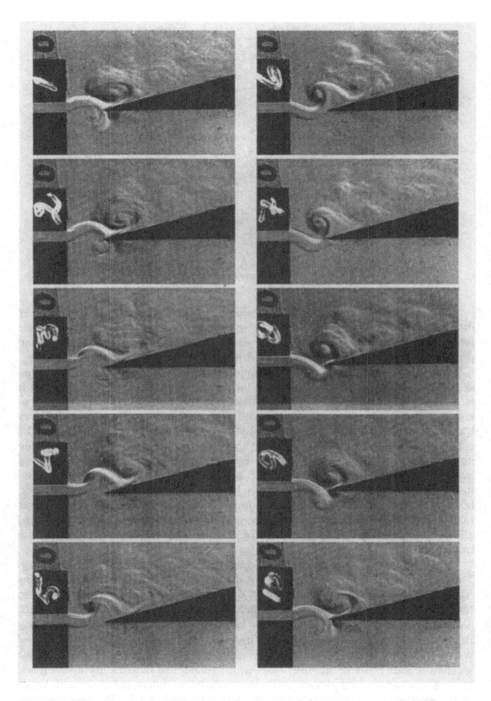

Fig. 6.5 Oscillatory flow in the mouth of a pipe organ. The sequence runs down the left-hand column and subsequently down the right-hand column.

instruments radiate only from their open ends. Termination by an expansion section (horn) allows designers to manipulate the frequencies and relative strengths of the harmonics and the directional radiation characteristics.

There has been considerable interest in the question of whether or not the vibrations of the walls of wind instruments contribute significantly to the radiated sound. Some manufacturers and players claim that the type of material from which an instrument is made does affect its sound quality. There may be a small influence, but the walls of small diameter cylinders are very stiff, and the amplitudes of vibration excited by the sound pressures in the air inside wind are extremely small.

6.3   PNEUMATIC TYRES

At first thought it seems obvious that it is the air pressure in the tyres that supports road vehicles. Further consideration suggests that this may not be the whole story. Consider first the case of a person sitting on a large inflated rubber ball, in which the air pressure directly produces forces on the ground and on the rider's body through contact with the thin flexible wall of the ball (in addition to some force due to the distortion of the spherical surface by the weight of the rider, which can add some tension force to resist the imposed weight). The surface of contact with the rider's body has no equivalent in the case of a tyre because the weight of the car is transmitted to the tyre via the contact between the tyre wall and the rim, not via air pressure. The pressure within the tyre, that acts on both the tyre and the rim, is uniform; so there is no net pressure force on the rim. Place the wheel upright on the ground. The tread in contact with the ground is slightly squashed and the lowest part of the tyre wall bulges a little more; but the pressure is still uniform. Now put the tyre on a car. The tyre wall bulges more, the area of contact with the ground increases, but the pressure is still uniform and exerts no net force on the rim. Reduce the tyre pressure and the lower part of the wall bulges more, and ultimately buckles out of shape and the contact patch increases in size, but there is still no net pressure force on the rim and the wheel still supports the car. Clearly, the force on the rim must be produced by the tyre carcass. But this comprises a wall made of thin flexible rubber reinforced with steel wires and fabric. So, how does it do it?

The function of the air pressure is to press outwards radially on the tyre belt so that it resists deformation from its circular shape and in so doing puts the side walls into tension, which can therefore act in a manner roughly similar to the spokes on a bicycle wheel. The inner boundaries of the tyre walls are kept in strong contact with the rim by air pressure. When the axle load is placed on the rim, the rim moves down and friction between the boundaries and the rim takes the inner boundary with it, thereby distorting the tyre shape. This deformation is resisted by air pressure and by elastic stresses set up in the tyre wall. The air pressure in a pneumatic tyre stiffens the tyre wall but it does not actually support the weight placed on the axle.

6.4   AIR BEARINGS , PALLETS AND CASTERS

Mechanical bearings are widely used to transfer loads from rotating shafts and other moving structures to their support structures. They normally incorporate either ball or roller elements running between inner and outer races and lubricated by grease or other forms of liquid lubricant. One of the disadvantages of these elements is that they contact the surfaces of the races over very small areas, which produces high stress concentrations that ultimately lead to degradation of the surfaces of both the elements and the races. Non-uniformities of element shape caused by wear lead to excessive vibration and noise and generate a lot of heat, and lubricants can become contaminated with debris from damaged elements.

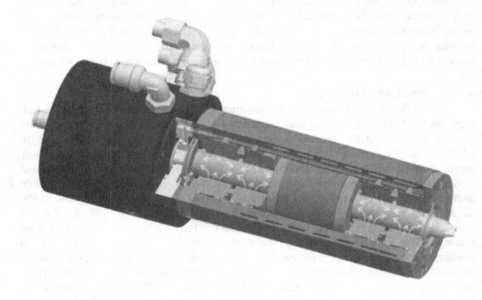

Fig. 6.6 An air bearing

An alternative technology employs air as the load bearing and lubricating medium. In air bearings, pressurised air is pumped into a very thin annular space between the rotating and fixed components, as illustrated by Fig. 6.6. There is no physical contact. Air bearings are particularly effective for high rotational speeds and have run at up to 360,000 revs per minute (6000 revs per second). They maintain the axis of rotation with great precision because they automatically smooth out any imperfections in the bearing surfaces and the air layer is stiffer than the equivalent mechanical bearing because the load is distributed evenly over the bearing surfaces rather than at very localised contact points which undergo local (Hertzian) deformation. They produce very little vibration or noise and are cooled by the air that slowly leaks through the bearing gap. Air bearings are widely used in computer disc drives.

Air bearing tables are used to allow machinery components to glide smoothly across each other and to be moved and positioned extremely precisely, which is of particular importance in the manufacture, assembly and inspection of products such as semi-conductor equipment, optical fibres and photonics devices. Another form of air bearing is the hover-pallet that is employed to allow heavy loads to be easily moved across fairly smooth surfaces such as factory or warehouse floors. They take various forms, but share common features. In one form a continuous, airtight cushion is slipped under the periphery of the pallet which carries the load to be moved. The cushion is inflated to raise the pallet off the ground and then the resulting cavity is fed with compressed air. The pressure supports the pallet and the leakage flow under the cushion 'lubricates' the interface with the ground to produce very low friction. Many tons can be moved over a smooth horizontal surface by one man. A heavy duty aero-caster platform carrying a 60 ton transformer is illustrated in Fig. 6.7.

Fig. 6.7 Air-supported load-transporting platform

The read/write heads of computer hard discs 'float' on an extremely thin layer of air that is maintained by the relative motion of the head and the disc. The geometric forms of the heads are critical to the accurate tracking and location of the heads. The mechanism of this form of aerodynamic lift is similar to that exhibited by a sheet of paper that is released so as to slide closely above a smooth surface such as a table.

6.5   FLUIDICS

Fluidics, also known as Fluid Logic, is the technology of manipulating fluid flows to create binary switches and amplifiers so that fluidic devices can be used to construct control circuits and analogues of other electronic circuits and devices. Fluidic amplifiers and switches are principally based upon the fact that, under some conditions, jets of fluid are sensitive to disturbance by much weaker jets that impinge on them from a lateral direction. An example is shown in Fig. 6.8. Fluidic devices have certain advantages over their electronic counterparts. They are rugged and do not carry electric charge, so they can be employed in situations where electronic devices would either be vulnerable to damage, as under conditions of extreme vibration or noise, or are potentially dangerous, as in areas containing flammable or explosive gases. An advantage that is of particular interest to the military is that they are not vulnerable to interference by electromagnetic fields.

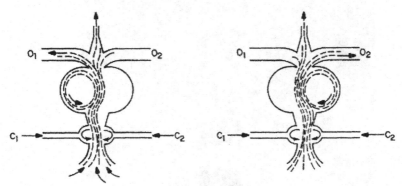

Fig. 6.8 A fluidic control circuit

Fluidic control devices are currently being used in unmanned aerial vehicles (UAVs). They are also finding use in the rapidly growing field of nanotechnology, which involves the manipulation, transport, and distribution of materials on molecular and cellular scales. Some fluidic devices are designed to physically control the distribution of liquids, such as windscreen washers and spa nozzles. The principal shortcoming of fluidic-based control devices is that their response times are rather slow compared with those of electronic devices. However, there are many applications in which very rapid response is not required.

6.6   SHIP DRAG REDUCTION

The drag of ships and boats has three principal components. The bow and stern waves, together with smaller waves radiating from the sides of the boat, carry away energy. Pressure drag arises from failure of the surface flow to close smoothly behind the vessel in much the same way as that described in Chapter 3. Skin friction drag arises from the action of viscous stresses on the hull associated with the boundary layer; this is the largest component of drag for most conventional boats and ships. The skin friction drag can be reduced by using air in three different ways. Microbubbles can be injected from the hull into the water to from a 'carpet'. This does reduce skin friction drag to some extent, but it is really only suitable for slow, flat-bottomed, cargo ships because it is virtually impossible to stop the bubbles from escaping any surface that is not horizontal. At higher speeds, the boundary layer turbulence tends to drive the bubbles away from the surface and the entrainment of bubbles into flow approaching the propellers reduces their propulsive efficiency. Another method is to trap a film of air between the hull and the water by pumping air across a highly water repellent coating.

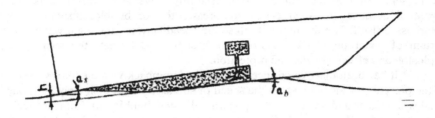

Fig. 6.9 Air Cushion Ship concept

However, the most effective design, which has been adopted by a number of commercial ship designers, is the air-cavity ship (ACS). Air is injected into a cavity which separates a substantial portion of the bottom of the hull from the water, as shown by Fig. 6.9. The pressurised air supports that area but the tangential interface stresses are far less than those between the water and a solid surface: the skin friction drag is therefore greatly reduced. The concept was first suggested by William Froude and Gustaf de Laval in the 19th Century. The practical difficulties of modelling, analysing and optimising the complex air-water interaction have delayed practical implementation until fairly recently. Current designs reduce drag by up to 40%, with a penalty of about 3% incurred by the necessary pumping power. The ACS is different from the hovercraft (or air cushion vehicle) because the cavity air pressure does not fully support the weight of the vessel and the required pumping power is very much less. The latter can, of course, also effect the transition from water to land. The early practical development of the ACS concept took place principally in Leningrad (now St. Petersburg) at the Krylov Research Institute. The application of the concept to river barges demonstrated drag reductions of up to 30%. Although ACS vessels are fairly common in Russia, economic weakness and military secrecy have, until recently, hindered their spread to the rest of the world. Alternative, but related, concepts are now being developed by other technologically advanced nations.

6.7 BUBBLE CURTAINS

As explained in Chapter 4, sound involves an interaction between the elasticity (or stiffness) and mass of the supporting medium. Sound in water travels faster than sound in air because its stiffness is 15,000 times that of air, but its density is only 800 times that of air. One of the techniques used to create, or enlarge, underwater channels in a rock seabed is the use of controlled explosions, as in quarries on land. The very powerful sound waves generated by these explosions are capable of damaging underwater structures in the vicinity and also pose a very serious danger to any divers or sea creatures within a large radius of the explosion. In order to minimise such undesirable effects, 'curtains' of air bubbles are used to screen the explosions. It is a remarkable fact that a 1% concentration of air bubbles in water reduces the speed of sound from about 1450 m/s to about 120 m/s. The inertial contribution of the water to sound wave propagation is not greatly altered by the presence of bubbles, but the compressibility (inverse of stiffness) is greatly increased because the water easily compresses the air bubbles instead of being compressed itself. The discontinuity in sound speed between the plain water and the curtain of water populated with bubbles scatters and reflects the sound of the explosion and effects the desired protection.

It has been discovered recently that the humpback whale exploits the effect of the presence of air bubbles on the speed of sound to corral their prey. A group of whales swims round in a circular pattern releasing bubbles to form an acoustic 'cage', as seen in Fig. 6.10(a). They swim around the cage making loud 'trumpeting' noises. These sounds are trapped by refraction in the walls of the 'cage'. The fish are afraid to leave the cage because of the noise in the wall of sound and are trapped, making them easy meat for whales rising from below and birds from above, as seen in Figs. 6.10(b) and (c). It is believed that bubble generation also plays a role in the fishing strategies of other marine hunters.

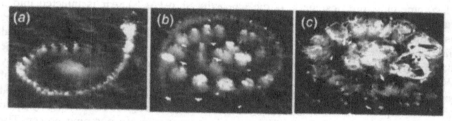

Fig. 6.10 Humpback whales corraling their prey

6.8   AIR SPRINGS

We have seen in earlier chapters that air trapped in closed volumes can be extremely stiff. This property is exploited in air springs, in which air is enclosed in a flexible membrane. These provide the particular advantage over mechanical springs that the structure that they support can be easily raised or lowered. You will almost certainly have observed this capability as demonstrated by the suspensions of buses that can be lowered to ease the access of passengers. They are also installed in some types of all-terrain cars of which the ground clearance must be increased when traversing extremely uneven terrain. This capability also allows self-levelling suspension systems to be constructed. The stiffness of air springs can be quite rapidly changed by charging or discharging air. This facilitates manual or automatic control of vehicle suspension stiffness. Air springs can provide active control of vibration to which precision machine tools and industrial inspection systems are subjected, either by their internal activity, or by their supporting structure – for example by the operation of other nearby machines or passing heavy road or rail traffic. Air springs are also widely used on railway vehicles.

Another application of the springiness of air is in the control of the transmission along piping systems of the unsteady pressures generated by liquid pumps, which can otherwise damage pipework components and generate excessive vibration and noise. Every one knows that expansion chambers (mufflers) are fitted to the exhaust pipes of internal combustion engines to reduce the noise radiated to the atmosphere. The principle of operation is that the sound waves generated when the exhaust valves open and close are reflected by the sudden changes of pipe cross section at the entry and exit of the chamber. This effect is similar to that which reflects sound at the open holes of wind instruments. It is not generally practicable to fit such a simple device to pipes carrying liquids such as oil that is pumped over long distances. Instead, a section of pipe is fitted with expansion section filled with air or other gas, compressed to match the local liquid pressure. The gas is separated from the liquid by a flexible coaxial membrane. The compressibility of the gas is far greater than that of the liquid. When a sound wave meets the expansion section, the sudden release of oscillatory pressure strongly reflects the sound.

6.9   COMPRESSED AIR TOOLS
We are all uncomfortably familiar with pneumatic drills in which a piston is repeatedly fired against the top of a drill bit by blasts of compressed air. It is, perhaps, not so well known by the general readership that compressed air is used to drive many different types of tool, particularly when employed in conditions where electricity is potentially dangerous. Examples include drills, riveters, chippers, chisels, caulking hammers, grinders, cutting tools, polishers, sanders, ratchet wrenches and screwdrivers. Air-driven tools are light, do not overheat and require little maintenance. High pressure air jets are widely used for cleaning and removing accretions from surfaces. Compressed air is used for spray painting and sand blasting. Sculptors often use air-driven chipping and scaling hammers. Fire suppression sprinkler systems are controlled by air pressure which prevents water from being fed into the pipes until the pressure is released when heat breaks a seal.

Turning to a very unpopular subject, modern dental drills are driven by air turbines, some of which rotate at up to 800,000rpm (13,333 rps): but 200,000 rpm (3333 rps) is more common. This frequency is in the most sensitive range of human hearing, which contributes to the fear that some patients suffer in the dentist's chair. The original air-driven drill was invented and patented in New Zealand in 1949. It was developed into commercial form in the US and first professionally adopted in 1957. Dental drills are also used by glass engravers and many other craftsmen for precision work. An air-driven abrasion device (like small sand blaster) is sometimes used to remove unwanted tooth material.

6.10   SEWAGE TREATMENT
Modern packaged sewage treatment plants are suitable to serve small groups of houses or small commercial premises. They are designed to encourage naturally occurring bacteria within the sewage to form a 'biomass' which breaks down the sewage material and cleans the liquid component. The process involves the introduction of air bubbles to support the aerobic bacteria and to stir up the material in what is known as the 'agitated sludge' method

6.11   PNEUMATIC TRANSPORT OF MATERIAL
Pneumatic conveyors use air to transport materials from place to place by means of pressure differences/gradients generated in pipework. Very dense and abrasive materials are moved by generating higher-than-atmospheric pressure to 'push' them; low density materials are generally moved by generating less-than-

atmospheric pressure to pull them along. The range of materials handled includes powders, grains, soils, sands, wood chips, toxic waste and asbestos. Industries that employ pneumatic conveyors include food and drink processing, plastics, pharmaceuticals, machining, powder coating, clay and ceramics, glass, mining, paper and pulp, and steel and chemical. A major advantage of pneumatic conveyors are that they are sealed, therefore preventing both ingress and egress of material, which is very important for the safe handling of toxic materials and for the hygienic handling of foodstuffs. Pneumatic conveyors are also used to convey bagged materials in bulk.

6.12   AIR CUTTING

The nature of plasma was described briefly in Chapter 1. An ionised gas contains atoms that have had electrons stripped off them. This allows the gas to become a good conductor of electricity, which is the transport of electrons. The plasma arc process was developed in the 1950s out of the earlier development of electric arc welding during the Second World War specifically for the purpose of aircraft production. In the latter process, a high voltage electric arc was formed between a metal rod and the work piece. In order to prevent oxidation of the molten metal, the surrounding air was excluded by means of blowing an inert gas, such as argon or helium, to form a shield around the arc. Engineers attempting to improve the process found that by reducing the gas nozzle opening, the electric arc and gas flow were constricted, the arc temperature and voltage greatly increased, and the resulting plasma cut the work piece instead of welding it. Many different plasma cutting techniques and gases have since been developed. One of the most popular is Low-Amp Air Plasma Cutting (Fig. 6.11) which can be applied by lightweight units supplied with currents of less than 200 amps.

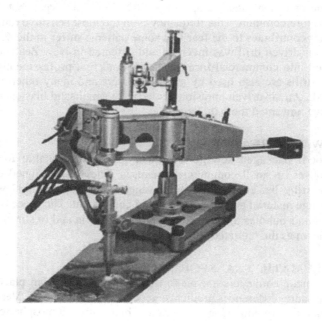

Fig. 6.11 Low –Amp air plasma cutting system

It has been found recently, that the use of compressed air to assist laser cutting of metal sheets is cheaper than the current use of nitrogen and oxygen, as well as facilitating faster cuts. The compressed air creates a plasma ball at the surface of the material, which cuts the metal. If sufficiently high pressure is used, it is possible to cut 12 mm thick aluminium sheet with a 4 kW laser resonator.

6.13   AIR BAGS

Ironically, air bags in automobiles are not inflated by air. A pyrotechnic action is triggered by the signal from an accelerometer that senses the impact. This generates gas (usually nitrogen) extremely rapidly (typically within 50 milliseconds) which inflates the folded bag.

Fairwell

I hope that you enjoyed this book, or at least, parts of it. If you have any comments or questions that you would like to put to me, please contact me at frank.fahy@gmail.com. I would very much welcome your feedback – both supportive and critical!

Some suggestions for further reading

The Atmosphere
An Ocean of Air. Gabrielle Walker. Bloomsbury, UK. 2007. ISBN: 978-0747581901

Global Warming
The Hot Topic. Gabrielle Walker and Sir David King. Bloomsbury, UK. 2008. ISBN: 978-0747593959

Oxygen
Oxygen: The Molecule That Made the World. Nick Lane. OUP, UK. 2003. ISBN: 978-0198607830

Fluid Flow
An Album of Fluid Motion. Milton Van Dyke. The Parabolic Press, Stanford, California. 1982. ISBN: 0-915760-02-9

http:// web.mit.edu/hml/ncfmf.html   Excellent movies about fluid dynamics

A History and Philosophy of Fluidmechanics. G A Tokaty. G T Foulis & Co, Ltd, Henley-on-Thames, UK. 1971. ISBN: 0854291180. [Out of print but worth seeking in libraries]

Aircraft and Flight
Aircraft Flight: A Description of the Physical Properties of Aircraft Flight. R H Barnard and D R Philpott. (3rd Edition.). Prentice Hall. 2003. ISBN:978-0131200432

Bird and Insect Flight
Nature's Flyers: Birds, Insects and the Biomechanics of Flight . David Alexander. The John Hopkins University Press. 2004. ISBN: 978-0801880599

Avian Flight (Oxford Ornithology Series). John Videler. OUP, USA. 2006. ISBN: 978-0199299928

Sound and Music
Exploring Music: The Science and Technology of Tones and Tunes. Charles Taylor. Taylor and Francis. 1992. ISBN: 978-0750302135

Pneumatics
Practical Pneumatics. Chris Stacey. Butterworth-Heinemann, UK. 1997. ISBN: 978-0340662199

Meteorology
Coriolis phenomenon
www.whoi.edu/science/PO/people/jprice/class/aCt.pdf (rigorous and mathematical)

Credits

Numbers indicate figures
*indicates that no reply has been received to the permission request or
source unknown

Cover illustration: from NASA Langley Research Center:
Photo ID: EL-1996-00130

1.1   Reproduced with the kind permission of Michael Wößner:
www.kowoma.de

1.2   Image Science and Analysis Laboratory, NASA-Johnson Space Center,
The Gateway to Astronaut Photography of Earth

3.3(a)   Reproduced with the kind permission of Gale M Craig from 'Introduction
to Aerodynamics', Regenerative Press, Anderson, IN, USA, 2002

3.3(b)   Photo by Thomas C Corke* in 'An Album of Fluid Motion' ed. Milton Van
Dyke, The Parabolic Press, CA, USA, 1982

3.4   Reproduced from 'An Informal Introduction to Theoretical Fluid
Mechanics' , James Lighthill, Clarendon Press (OUP), UK, 1986

3.5   Photo by D H Peregrine* in 'An Album of Fluid Motion' ed. Milton Van
Dyke, The Parabolic Press, CA, USA, 1982

3.6   Reproduced from 'Mechanics of Fluids', W J Duncan, A S Thom and A D
Young, Edward Arnold, London, UK, 1970

3.7   Reproduced from 'Applied Hydro- and Aeromechanics', L Prandtl and O G
Tietjens, McGraw-Hill Books Inc, New York, 1934

3.8   As 3.7

3.9   Photo by Daimler-Chrysler reproduced from 'Autospeed', Issue 438,2007,
Web Publications Pty Limited*

3.10   As 3.6

3.11   Reproduced from 'An Introduction to Fluid Mechanics' , G K Batchelor,
Cambridge University Press, UK, 2000

3.12   Reproduced from 'Complex Analysis for Mathematics and Engineering',
John H Mathews and Russell W Howell, Fifth Ed., 2006, Jones and Bartlett
Publishers, Sudbury, MA, USA, www.jbpub.com

3.13   As 3.7

3.14   Source unknown*

3.15   Photo by Joel Vogt* reproduced from www.airliners.net

3.16   As 3.3(a)

3.18(a)   Image courtesy of NASA Glenn Research Center

3.18(b)   Photo by the US Air Force: see http://www.af.mil/weekinphotos/060526-
03.html

3.19   Photo by Sadtoshi Taneda* in 'An Album of Fluid motion' ed. Milton Van
Dyke, The Parabolic Press, CA, USA, 1982

3.20   Reproduced with the kind permission of the American Institute of Physics
from H Yamada & T Matsui, Phys. Fluids, **21**, 292, 1978

3.21   Reprinted from 'The Nature of Structural Design and Safety, David
Blockley, Ellis Horwood, Chichester, UK, 1980

3.23   Photo courtesy of Scott Rathbone

3.24   Photo reproduced from 'An Album of Fluid motion' ed. Milton Van Dyke, The Parabolic Press, CA, USA, 1982 with the kind permission of Thomas J Mueller

3.25   As 3.7

3.26   From the George Grantham Bain Collection of the US Library of Congress

3.27   Reproduced from the 'Annual Reviews of Fluid Mechanics', 17,1985 with the permission of Annual Reviews: www.annualreviews.org

3.28   Photo by the US Navy: see http://.www.news.navy.mil/view_single.asp?id=27252

3.29   Image courtesy of John J Videler (see suggestions for further reading)

3.30   Reproduced from ' A family of vortex wakes generated by a thrush nightingale in free flight in a wind tunnel over its entire natural range of flight speeds', G R Spedding, M Rosén and A Hedenström, The Journal of Experimental Biology, 206, 2313-2344

3.31   Photo courtesy of Roy J Beckemeyer

3.32   Photo courtesy of John Altringham, University of Leeds, UK

3.33   Image reproduced from 'In the Beginning….. The SR-N1 Hovercraft', R L Wheeler and J B Chaplin, Crown Publishing, Isle of Wight, UK

3.34   Image reproduced from http://www.se-technology.com/wig/index.php*

3.35   As 3.34

4.3   Images reproduced from 'Noise and Vibration Control Engineering', eds. L L Beranek and I Vér, John Wiley and Sons, New York, 1992

4.4   As 4.3

4.5   Reproduced from 'Fundamentals of Noise and Vibration', eds. Frank Fahy and John Walker, E & F N Spon, London, 1998

4.6   Photo courtesy of Gary S Settles, Penn State University, USA

4.7   Out of copyright

4.8   Reproduced from http://en.wikipedia.org/wiki/Turbofan *

5.2   Reproduced from The Free Dictionary, Helicon Publishing, Research Machines plc 2009

5.3   Reproduced from http://en.wikipedia.org/wiki/Coriolis

5.4   Reproduced from http://visibleearth.nasa.gov/view-rec.php?id=6204

5.5   Reproduced from http://www.ecmwf.int/research/era/ERA-40/ERA-40-Atlas/images/full/DO6_LL_JJA.gif

5.6   Reproduced from weather.com: The Weather Channel Interactive*

5.7   Reproduced from http://www.fortlauderdale.gov/police/hurricane/images/hurricane.jpg*

5.8   From web material from NOAA Central Library; OAR/ERL/National Severe Storms Laboratory (NSSL), USA

5.9   As 5.9

6.2   Reproduced from 'Acoustics and the Performance of Music', Jürgen Meyer, PPVMedien, Germany

6.3   Reproduced from 'The Physics of Musical Instruments', Neville H Fletcher and Thomas D Rossing, 1991, with the kind permission of Springer Science and Business Media

6.4   As 6.3

6.5   Reproduced from 'Vortex shedding on steady oscillation of a flue organ pipe', B Fabre, A Hirschberg and A J P Wijnands, Acustica with acta Acustica **82**, 863-877, 1996 with the permission of Medipharm Scientific Publishers – S. Hirzel Verlag, Stuttgart, Germany

6.6   Image courtesy of Westwind Air Bearings, a Division of the GSI Group Ltd, UK

6.7   Reproduced from www.aerogo.uk

6.8   Reproduced from US Patent # 4,000,757

6.9   Reproduced from http://docs.hydrofoils.org/SAS03.pdf in Speed at Sea/Feb/2003*

6.10   Reproduced from 'Cavitation and Cetacean', Timothy G Leighton, Daniel C Finfer and Paul White, Revista de Acustica **38**, 3-4, c. Tim Voorheis*

6.11   Reproduced from http://www.germes-online.com/catalog/17/739/71216/inverter_dc_air_plasma_cutting_machine.html c Shaoxing Yuejian Machinery Co Ltd, Canton, P R China

Chapter 3 Epigraph: Poem by John Gillespie Magee. Reproduced from David English, *Slipping the Surly Bonds*: *Great Quotations on Flight*, McGraw-Hill, 1998.

Index

Printed in the United States
By Bookmasters